Springer Series in Reliability Engineering

For further volumes:
http://www.springer.com/series/6917

Cher Ming Tan · Zhenghao Gan ·
Wei Li · Yuejin Hou

Applications of Finite Element Methods for Reliability Studies on ULSI Interconnections

 Springer

Dr. Cher Ming Tan
School of Electrical and Electronic
  Engineering
Nanyang Technological University
50 Nanyang Avenue
Singapore 639798
Singapore
e-mail: ecmtan@ntu.edu.sg

Dr. Zhenghao Gan
Technology Research and Development
Semiconductor Manufacturing
International
  (Shanghai) Corporation (SMIC)
18 Zhangjiang Road
Pudong New Area
201203 Shanghai
People's Republic of China
e-mail: Howard_Gan@smics.com;
    ezhgan@163.com

Dr. Wei Li
Singapore Institute of Manufacturing
  Technology
71 Nanyang Drive
Singapore 638075
Singapore
e-mail: wli@SIMTech.a-star.edu.sg

Dr. Yuejin Hou
#10-226 BLK156
11 Hougang Street
Singapore 530156
Singapore
e-mail: HOUY0001@e.ntu.edu.sg

ISSN 1614-7839

ISBN 978-0-85729-309-1   e-ISBN 978-0-85729-310-7

DOI 10.1007/978-0-85729-310-7

Springer London Dordrecht Heidelberg New York

British Library Cataloguing in Publication Data
A catalogue record for this book is available from the British Library

Cover design: eStudio Calamar, Berlin/Figueres

Printed on acid-free paper

Springer is part of Springer Science+Business Media (www.springer.com)

Contents

Chapter 1
Introduction

An integrated circuit is made up of transistors and interconnections, and hence the reliability of an integrated circuit depends on the reliabilities of transistors and interconnections. As the interconnect line width is shrinking in ULSI, the reliability of the interconnection is deteriorating significantly, if nothing is done for the interconnection design and process; rendering the increasing importance in the reliability study of interconnections. In fact, study has shown that the failure rate associated with interconnect failure in ULSI is increasing drastically as the technology node advanced [1]. Further elaboration on the importance of interconnection reliability in ULSI can be found in a recent book entitled "Electromigration in ULSI" published by World Scientific [2].

In today's ULSI interconnection system, three failure mechanisms are typical, namely the electromigration, stress induced voiding and low-k dielectric breakdown. This is also spelt out in the ITRS roadmap 2009 in the Interconnect Chapter. With the increasing complexity of the interconnection system and its 3D nature in its actual implementation, together with the use of low-k dielectric, the factors that affect the interconnection reliability are no longer simply current density, and the temperature distribution as well as the resulting thermo-mechanical stress distribution are also important influencing factors that must be considered [2]. This renders the study of interconnection reliability difficult. While one can perform reliability test such as electromigration test, stress-induced voiding test and time-dependent dielectric breakdown (TDDB) test, the test time can be too long to meet the development time requirements in semiconductor industry, and also the underlying mechanisms that are interacting during the degradation process are difficult to identify. However, such identification of the dominant failure mechanisms and their relationship with the interconnect structures, materials, and process are crucial in the design-in reliability for integrated circuit.

Facing the challenges in the reliability study of interconnection system as mentioned above, the most vital solution is to employ physics-based simulation

C. M. Tan et al., *Applications of Finite Element Methods for Reliability Studies on ULSI Interconnections*, Springer Series in Reliability Engineering, DOI: 10.1007/978-0-85729-310-7_1, © Springer-Verlag London Limited 2011

and modeling. As most physical systems can be described using a set of partial differential equations, finite element method (FEM) has evolved to be a good tool in solving these partial differential equations and obtain solutions that represent the physical process of degradation of interconnection system in ULSI. As our understanding of the failure mechanisms of interconnection system increase with the help of FEM and experimentation, and the increasing power of numerical computation hardware and software, models formulated using FEM can be complex and their results are becoming closer to the reality. Hence, key influencing factors on ULSI interconnection reliability can be identified using FEM, and one can then design an interconnect system appropriately for desired reliability.

Besides the identification of the failure mechanisms of interconnection system, the determination of the effective k-value of a low-k interconnect system and the effect of process defect on the effective k-value can also be determined using FEM. This is important because the complex interconnection structure precludes us to compute the effective k-value analytically. It is always good to perform the calculation of the effective k-value in an interconnection system before the actual fabrication because the fabrication cost for today's ULSI is too expensive, for not doing it right the first time.

In the past few decades, many research works were done to improve the modeling capability of the interconnection system, and significant progress has been made. In this book, we attempt to put together the various works on the modeling of the reliability as well as the determination of effective k-value of a low-k interconnection system. FEM is traditionally used for mechanical engineering, and this book is the first book in the description of FEM in the reliability study in semiconductor industry.

The structure of the book is designed to lead the readers working or interested in ULSI interconnection from the basic understanding of the failure mechanisms of interconnection and the FEM to the step-by-step application of FEM in interconnect reliability modeling. The progress of the modeling on interconnection reliability is also introduced so that readers can understand the rationale behind the various considerations in the FEM today. The determination of the effective k-value of an interconnection system and the examination of the various process defects on the effective k-value with the aid of FEM will be presented in the last chapter.

Physics-based FEM for ULSI interconnection system is a complex subject. While significant progress has been achieved in recent years, there remains a vast research opportunity to improve the modeling capability as spelled out in the recent book on electromigration [2], and this calls for international close collaboration to advance the FEM. Unfortunately, workers on FEM for interconnection are mainly working individually within their groups, and the support from industry in providing data needed for modeling is weak, unless the modeling group is within the company themselves. This can slow down the progress of the FEM development and duplicate effort, and can indeed work its way out of the industry as interconnect technology is changing rapidly and new interconnect materials may be replacing the Cu, rendering limited successful applications of the FEM in interconnection system design and process improvement, to achieve the design-in reliable interconnection system.

References

1. Srinivasan JA, Bose P, Rivers JA (2004) The impact of technology scaling on lifetime reliability. In: 2004 International Conference on Dependable Systems and Networks, pp 177–186
2. Tan CM (2010) Electromigration in ULSI interconnection. World Scientific Publishing Co

Chapter 2
Development of Physics-Based Modeling for ULSI Interconnections Failure Mechanisms: Electromigration and Stress-Induced Voiding

As mentioned in Chap. 1, the two major failure mechanisms in ULSI interconnections are electromigration and stress-induced voiding. In this chapter, we will provide reviews on the model development for electromigration and stress-induced voiding.

2.1 Electromigration (EM) Modeling Review

In this chapter, we present a comprehensive review on the physics-based modeling of EM phenomena in ULSI interconnections over the last three decades. In the evolution of the physics-based modeling, some aspects of the physics are dropped for simplification, and some are added to accommodate new understanding on the EM physics as well as for the new development of the interconnect technology. With the continuous change in the metallization system and materials, the aspects of physics that have been dropped may become important again, and new physics might also occur with these changes in metallization system. Here, we re-examine the justification of dropping or adding various physical aspects in the EM modeling during their evolution and their implications on the future interconnect system.

2.1.1 One-Dimensional (1D) Analytical Modeling

Like most physical modeling, physics-based EM modeling began from 1D analytical modeling by solving the following continuity equation [1] derived from the time evolution of the vacancy concentration along an interconnect line:

C. M. Tan et al., *Applications of Finite Element Methods for Reliability Studies on ULSI Interconnections*, Springer Series in Reliability Engineering, DOI: 10.1007/978-0-85729-310-7_2, © Springer-Verlag London Limited 2011

$$\frac{\partial C_v}{\partial t} + \frac{\partial J_v}{\partial x} + r = 0. \tag{2.1}$$

Here C_v is the instantaneous vacancy concentration at position x and time t, J_v is the vacancy flux due to EM-driving forces and r is the sink/source term which allows for the recombination or generation of vacancies at sites such as grain boundary, dislocations, or surface [1].

The focus of this 1D modeling is on the simulation of EM kinetics. The net vacancy flux along the length of an interconnect can be due to the electron wind force, vacancy diffusion as a result of concentration gradient (Fickian diffusion) and temperature gradient (Soret diffusion), hence the flux can be written as [1]

$$J_v = -D\frac{\partial C_v}{\partial x} + \frac{DC_vZ^*eE}{k_BT} + \frac{Q^*DC_v}{k_BT^2} \cdot \frac{\partial T}{\partial x} \tag{2.2}$$

where D is the vacancy diffusivity given by $D = D_0 \exp(-\Delta E_a/k_BT)$, E_a is the activation energy for the vacancy diffusion, Z^*e is the effective charge of the diffusing species, E is the electric field and Q^* is the heat of transport.

The sink/source term r can be expressed as [1]

$$r = \frac{C_v - C_{ve}}{\tau} \tag{2.3}$$

where C_{ve} is the equilibrium concentration of vacancies within a grain, and τ is the average lifetime of a vacancy.

Equation 2.1 is a basic EM equation from the first principle, and it was adopted by many 1D analytic EM models reported in the literature as will be discussed below. Various EM models were proposed based on different boundary conditions, and the phenomenon of EM were simulated by analyzing the evolution of the vacancy concentration in an interconnect metal line using Eqs. 2.1–2.3. In general, two categories were being considered in this 1D analytical modeling as follows.

2.1.1.1 Category #1: 1D Modeling Without the Consideration of Stress Gradient-Induced Migration

In one of the pioneer works on 1D continuum EM modeling, the work of Rosenberg and Ohring [2], the electron wind force, concentration gradients (Fickian diffusion), and temperature gradients are taken into consideration with the presence of the vacancy sink/source. As shown in Fig. 2.1, they proposed a model from the appropriate solution of Eq. 2.1 where it was assumed that two grains intersect at $x = 0$ and that each grain can be characterized by their respective vacancy diffusion parameters, namely the diffusivity D and the equilibrium vacancy concentration C_{ve}, the lifetime of the vacancy τ, and field strength E. The vacancy behavior as

Fig. 2.1 Model of two grain boundaries intersecting at $x = 0$. Reprinted with permission from Rosenberg and Ohring [2], copyright © 1971, American Institute of Physics

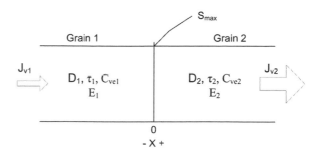

expressed by Eq. 2.1 was written separately for the two grains, assuming the diffusion parameters are constant within each grain.

In order for vacancy buildup to take place, it is necessary that the vacancy flux induced by the electron flow is different in the two grain regions. The boundary conditions used by Rosenberg and Ohring [2] were that the vacancy concentration and flux at $x = 0$ are continuous, and all the diffusion parameters are the same in the two grains except that the activation energies for diffusion are 0.6 eV for $x > 0$ and 0.8 eV for $x < 0$. Using Laplace transform method, both transient time-dependent solution and steady-state solution of $C_v(t)$ were obtained. The impact of temperature on EM kinetics and the discontinuity in grain size which causes an abrupt change in the number of atomic diffusion paths on the EM performance of a metal thin film were studied in their steady-state solution.

Their model simulated the vacancy saturation due to the divergence of vacancy flux at the grain boundary with the consideration of multiple EM-driving forces. Although the temperature gradient-induced migration was taken into consideration during their initial modeling of EM, it was dropped for the sake of simplicity during the actual calculation. For the justification of the simplification, they stated that the interconnect length was the only dimension considered in their 1D EM model, and the current density was evenly distributed throughout the length; thus the temperature along the length should be a constant as metals are very good heat conductor. However, in actual interconnection in today's ULSI, the effect of the temperature gradient can no longer be ignored as revealed by 3D modeling which will be discussed in Chap. 4. Also, the choice of the boundary condition with infinite ends in their model renders the model incapable of catering the length effect of a confined metal thin film; which was discovered by Blech few years later [3, 4]. Due to this incapability, this type of boundary condition was not chosen for subsequent 1D continuum EM modeling.

A more realistic boundary condition was proposed by Shatzkes and Lloyd [5]. In their study, the case of vacancy flow in the region of a purely blocking boundary was examined as shown in Fig. 2.2.

This condition can occur in real conductors, for instance, at the boundary of a "bamboo" grain. Mathematically, the boundary conditions for their case can be expressed as

$$C(-\infty, t) = C_0 \tag{2.4}$$

Fig. 2.2 Schematic drawing of the semi-infinite grain boundary considered in the work of Shatzkes and Lloyd [5]

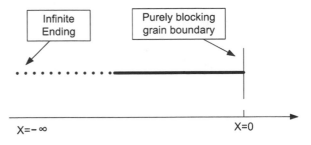

Fig. 2.3 Grain boundary between two triple points at $x = 0$ and $x = 1$. The angle between the grain boundary and the current density j is labeled as θ. Reprinted with permission from Kirchheim and kaeber [6], copyright © 1991, American Institute of Physics

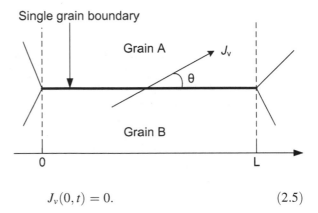

$$J_v(0, t) = 0. \qquad (2.5)$$

Although this solution treats a special case of a perfectly blocking boundary, it is useful for the understanding of the boundary conditions that allow a reduced vacancy flow as pointed out by them. In general, there is a flux through any given region, but a divergence may still exist. By treating this "residual" mass flux as a background, we can then treat the accumulation of vacancies above the background in the present manner. In their study, assuming isothermal condition with no thermal transport, the vacancy flux due to temperature gradient-induced migration was not considered as well. The vacancy sink/source term was also dropped in their EM equation because the formation and annihilation of vacancies in the grains can become significant only at elevated temperature [6].

Based on the same EM equation, Kirchheim and Kaeber [6] reported their EM modeling and solved the EM equation based on different boundary conditions. They considered the case of a single grain boundary structure as shown in Fig. 2.3.

Compared with the infinite boundary condition proposed by Rosenberg and Ohring [2] and the semi-infinite boundary condition proposed by Shatzkes and Lloyd [5], Fig. 2.3 shows a special case of finite boundary that the fluxes at both ends are zero (i.e., due to very low vacancy mobilities in the adjacent grain boundaries), that is,

$$J_V(x = 0) = J_V(x = l) = 0. \qquad (2.6)$$

In the work of Kirchheim and Kaeber, both the solutions with and without the consideration of vacancy sink/source were analyzed and discussed. Since the formation and annihilation of vacancies in the grains occur only at elevated temperatures as mentioned earlier, they focused on the steady-state concentration profile without vacancy sinks/sources. In the calculation of vacancy flux, the grain boundary component of the electric field or current density was considered, i.e., the total current density was multiplied by cos θ (refer to Fig. 2.3).

With this finite boundary condition, they simulated the mass backflow during EM. However, contrary to the generally accepted interpretation, where the mass backflow was a flow of matter caused by a stress gradient within the Al line [3, 4], their interpretation of the backflow was based on the diffusional backflow of vacancies with respect to the vacancy concentration profile. They argued that the vacancy fluxes and internal stress gradients are coupled since production and annihilation of vacancies is often accompanied by plastic deformation.

A more completed study on the boundary conditions of the EM equation was reported by Clement and Lloyd [7]. Similar to the previous studies, both the electron wind force and vacancy concentration gradient are taken into the consideration as driving forces of EM. In their report, the physical meanings of the three boundary conditions used were described and the solutions were compared at the blocking barrier. The first boundary condition corresponds to a situation of a semi-infinite metal line which is the same as the one proposed by Shatzkes and Lloyd [5]. The other two boundary conditions are for the cases of the finite interconnect length. The first case corresponds to a situation where the number of vacancies is conserved. An example of such case is an interconnect line with a thick and strong passivation layer that precludes volume change of the conductor, and such boundary condition and its corresponding solution have been reported by Kirchheim and Kaeber [6]. The second case corresponds to a situation where the passivation or adherent oxide layer on the metal line is not so strong and stiff as to preclude the creation of vacancies. This is what one might expect in an Al conductor covered only by its native oxide (Refer to Fig. 2.3). This new boundary condition can be expressed as:

$$J_v(L, t) = 0 \tag{2.7}$$

$$C(0, t) = C_0. \tag{2.8}$$

Solutions to the above-mentioned three reasonable boundary conditions for the EM equation were investigated numerically. It was shown that regardless of the boundary condition chosen, the time to failure would approximate the semi-infinite solution of Shatzkes and Lloyd as long as the critical failure vacancy concentration is significantly different from the vacancy concentration at equilibrium.

A complete review on the 1D EM modeling was contributed by Clement in his review paper [1]. He reviewed all the 1D EM models before 2001 from the perspectives of the consideration of driving forces of EM, choices of the boundary conditions, and the corresponding numerical solutions. In his report, the 1D EM models with the consideration of vacancy sink/source were compared to those without it. Although the numerical solutions of the EM equation for both cases are quite similar, there is one critical difference. The effect of the consideration of vacancy sink/source is the shift in time scale for the buildup of vacancies. Hence, the calculated time-to-failure will be in the order of several days, rather than a few seconds, as obtained from those models without the consideration. This highlights the importance of including the sink/source term in the 1D EM model. In his report, he also discussed the importance of critical components in 1D EM model, such as critical stress to initiate the void formation or dielectric cracking, critical void size to cause the failure of the interconnect, and current density dependence near the critical stress and void size. However, the aforementioned 1D EM overlooks the importance of stress effect during EM. Different hypothesis on the effect of stress during EM were proposed in the literature, and it was clear that the ignorance of stress effect in EM modeling would render inaccuracy in the prediction.

2.1.1.2 Category #2: 1D Modeling with the Consideration of Stress Gradient-Induced Migration due to Back Stress

With the shrinking interconnect line dimensions, it is found that, in the confined metal interconnects deposited on an oxidized silicon substrate and covered by a dielectric passivation layer, EM gives rise to back stresses that can retard EM itself [8]. A few reports on the modeling of EM incorporating the effect of the transient back stress buildup can be found, such as Ross [9], Kirchheim [10], and Korhonen et al. [11]. The modeling of the EM-driving forces was modified to account for the effect of the migration due to stress gradient.

For example, in the work by Kirchheim [10], additional EM-driving force, the stress gradient, was taken into consideration. Moving an atom from the grain boundaries to the surface changes the volume by one atomic volume, Ω. Allowing relaxation of the neighboring atoms leads to a contraction in volume of $f\Omega$ $(0 < f < 1)$, and hence the total volume change is $(1 - f)\Omega = f'\Omega$. With the presence of this strain, a stress gradient is produced, and the corresponding vacancy flux is given by

$$J_\sigma = -\frac{DC_v}{kT}f\Omega\frac{\partial\sigma}{\partial x}. \tag{2.9}$$

With this additional driving force, Eq. 2.2 was modified as

$$J_v = -D\frac{\partial C_v}{\partial x} + \frac{DC_vZ^*eE}{k_BT} + \frac{Q^*DC_v}{k_BT^2}\cdot\frac{\partial T}{\partial x} - \frac{DC_v}{kT}f\Omega\frac{\partial\sigma}{\partial x} \tag{2.10}$$

In their modeling, the atomic migration due to the temperature gradient was also neglected with the same argument as given by Rosenberg and Ohring [2]. Vacancy sink/source was considered in a similar way as in Eq. 2.3. The generation of a vacancy changes the volume by $\Delta V/V = f'\Omega$ and, therefore, the rate of the volume change within a grain of diameter d is related to the rate of vacancy generation within the grain boundary of thickness δ by

$$\frac{1}{V}\frac{\partial V}{\partial t} = f'\Omega\frac{\delta}{d}\frac{\partial C_v}{\partial t} = f'\Omega\frac{\delta}{d}\frac{C_v - C_{ve}}{\tau}. \tag{2.11}$$

Using the Hooke's Law which states that $\delta\sigma = B\delta V/V$, we have

$$\frac{\partial\sigma}{\partial t} = Bf'\Omega\frac{\delta}{d}\frac{C_v - C_{ve}}{\tau} \tag{2.12}$$

where B is the bulk modulus of the metal line.

With the two coupled differential equations (2.10) and (2.12), the continuity equation (2.1) for the vacancy concentration and the rate equation (2.3) for the vacancies sink and source reactions, the EM phenomenon was simulated in a finite line blocked at both ends $[J_V(x = 0) = J_V(x = l) = 0]$ as shown in Fig. 2.3. The incorporation of mechanical stresses developing as a consequence of electromigration and vacancy annihilation and generation, leads to a complicated relationship between vacancy concentration and mechanical stress which can only be solved analytically for a few limiting cases. By coupling stress development with the vacancy concentration change along the metal line, the stress gradient can be calculated and the back flow due to the stress gradient can be simulated.

In the model developed by Korhonen et al. [11], the driving forces of EM included only the electron wind force and the stress gradient, and both driving forces were formulated based on the atomic flux instead of the vacancy flux. However, the vacancy sink/source term was dropped in their consideration because they assumed that the net number of atoms entering a volume element was large enough to include vacancy sources and sinks. In their modeling, the change in atomic concentration was also coupled with the stress evolution in a similar manner as in the work by Kirchheim [10], and they presented several representative cases, such as the semi-infinite metal line where the flux is blocked at the end $x = 0$, and the confined finite metal line where the flux is blocked at both the ends. Besides the investigation of the EM equations with different boundary conditions, Korhonen et al. also studied the influence of the stress evolution on the diffusivity of atoms and its impact on the solutions to the EM equation. They derived the vacancy concentration as follows.

$$C_v = C_{v0}\exp(W_f + \Omega\sigma/kT) \tag{2.13}$$

where Ω is the atomic volume, W_f is the interaction energy between the vacancy and the stress field, and C_{v0} is the vacancy concentration in the absence of any stress effect. Additionally, the dependence of the vacancy diffusivity on the hydrostatic pressure p was also derived and included as follows.

$$D_v = D_v(0)\exp(-pV_m/kT) \tag{2.14}$$

where V_m is the dilatation associated with vacancy migration and $D_v(0)$ is the vacancy diffusion coefficient at $p = 0$.

The constant diffusivity used in the early solution (Kirchheim) was actually in the absence of any stress effects. The numerical integration of stress-dependent diffusivity in the EM model was accomplished by the method of finite differences. Surprisingly, numerical results with a stress-dependent diffusivity for the semi-infinite line turned out to be reasonably close to the analytic estimates with a constant diffusivity. Korhonen stated that the analytic solutions may in several cases be well sufficient for practical estimation of stress buildup during EM. However, for modeling of more complicated line structures, one must resort to numerical solution.

Although the EM equation formulated by Korhonen et al. [11] as discussed above is very similar to those by Rosenberg and Ohring [2], Shatzkes and Lloyd [5], Kirchheim and Kaeber [6], and Clement and Lloyd [7], the use of atomic flux with the consideration of the stress evolution by Korhonen et al. presented a retardation of EM damage due to the back stress. Such phenomena is not possible with vacancy flux approach as vacancy concentration gradient does not retard EM damage. Furthermore, significant vacancy concentration gradients can be created in a matter of seconds because of the almost negligibly small material transport involved. However, the stress evolution in the interconnect lines during EM can last hundreds of hours as predicted by Korhonen et al. [11] using the atomic flux approach.

With the development of the analytical modeling and interconnect technology, Korhonen's model received further review and continuous development in order to explain the physics of EM in advanced interconnect systems. Clement and Thompson [12] reported a complete review on the Korhonen's model and found that the analytic solution for a semi-infinite line with a blocking boundary given by Korhonen et al. was a good approximation only when the stress buildup is small, and this is usually not the case for narrow, encapsulated interconnect lines in which the EM-induced stress can be very high prior to failure. They proposed a complete model description and a more accurate analytic solution to the differential equations describing the EM-induced stress buildup at a blocking boundary [12].

Based on their improved model [12], Park et al. reported another EM model to simulate the reliability of Al and Al–Cu interconnects [13]. In their work, the effect of the impurity Cu atoms in Al interconnects on stress evolution and lifetime was investigated in various structures. In addition, the significance of the effect of the mechanical stress on the diffusivity of both the Al and Cu was determined. Current density exponents of both $n = 2$ for void nucleation and $n = 1$ for void growth failure modes were found in both pure Al and Al–Cu lines. More importantly, the application of the model was further extended to the investigation of other interconnection materials such as Cu and its alloys by modifying the input material properties. The detailed mathematical formulation used in the analysis can be found in their report and earlier study [14]. Their analytic EM model was developed into an EM simulation package called MIT/EmSim [15] which was extensively employed in their other EM studies [16–18] and successfully provided theoretical support to various experimental observations.

Although the stress gradient-induced migration due to back stress has been well addressed by the models discussed above, another important source of the stress is overlooked in their theoretical consideration, and it is the thermo-mechanical stress due to the thermal mismatch between the metal line and its surrounding materials.

The thermo-mechanical stress depends on the thermal expansion coefficients of the materials and the shape of the metal line itself; and 1D EM models are incapable of simulating the effect.

2.1.2 Two-Dimensional (2D) Modeling

Besides the electron wind force, vacancy concentration gradient, stress, and temperature gradients, the EM performance of an interconnect is found to be greatly affected by the microstructural inhomogeneities of the line, such as the grain size and texture distributions, the triple points of the grain boundaries, the barrier layer interface, and the surrounding materials of the metal line [19]. While the EM equation is generally easy to solve in one dimension, the simplifications fail to account for the details of the effect of microstructures. A step further into the reality is achieved by using 2D models with the consideration of the microstructure of the metal conductor. The addition of this second dimension complicates the governing EM equations significantly, and numerical approaches are generally required to find the solution. Most of 2D EM models use the atomic flux to study the EM instead of using vacancy flux. Some of them focus on the atoms transport along different diffusion paths, under various EM-driving forces as well as the void nucleation and growth, while some of them focus on the void shape evolution due to the atom transport along the void surface. Let us consider their respective models.

2.1.2.1 2D EM Driving Force and Diffusion Path Model

In 1999, Gleixner and Nix reported their study on the simulation of EM [20]. In their study, both the thermo-mechanical stress and electron wind force were taken into consideration. They proposed a new method of solution for the coupled stress-diffusion equations in a 2D model while avoiding the computational complexity associated with the finite element approaches. The atomic flux which results from electron wind force and stress gradients were expressed in Eqs. 2.15 and 2.16 respectively, as follows,

$$J_a = \frac{D}{\Omega kT} Z^* e \rho j \tag{2.15}$$

$$J_a = \frac{D}{kT} \nabla \sigma \tag{2.16}$$

where J_a is the atomic flux and $\nabla \sigma$ is the stress gradient. The stress is due to both the Blech effect σ_n and thermo-mechanical stress σ_h. The total atomic flux is then given by the sum of both equations as follows.

$$J_a = \frac{D}{kT}\left(\nabla\sigma + \frac{Z^*e\rho j}{\Omega}\right). \tag{2.17}$$

In their 2D model, they treated the mass transport along the grain boundary; along the sidewall interfaces and inside the grain differently.

For the case of atoms transport along the grain boundary, the atomic flux was given as:

$$J_a = \frac{D_{gb}}{kT}\left(\frac{\partial\sigma_n}{\partial l} + \frac{Z^*e\rho j\cdot\cos\theta}{\Omega}\right) \tag{2.18}$$

where D_{gb} is the diffusivity along the grain boundary, l is the distance along the grain boundary, and θ is the angle between current flow and the grain boundary.

For the case of atoms transport along the sidewall, the flux was given as:

$$J_a = \frac{D_{ini}}{kT}\left(\nabla\sigma_n + \frac{Z^*e\rho j}{\Omega}\right) \tag{2.19}$$

where D_{ini} is the diffusivity along the interface between the sidewall and the metal thin film.

For the case of atoms transport inside the grain, the flux was given as:

$$J_a = \frac{D_b}{kT}\left(\nabla\sigma_h + \frac{Z^*e\rho j}{\Omega}\right) \tag{2.20}$$

where D_b is the diffusivity inside the grain. The diffusion through the bulk is driven by a hydrostatic stress gradient instead of the normal traction due to the Blech effect. The stress field solution was calculated before the start of the simulation.

The stress field solution obtained was superimposed along all the diffusion paths to find the total stress field in the material. The local stress gradient and the diffusivities were then coupled in the calculation of local atomic flux. In this way, the method avoided the computational complexity associated with finite element approaches, and they claimed that this approach was able to simulate stress evolution in complex microstructures very quickly and can therefore be applied to interconnect with realistic grain structures.

Another 2D EM model based on Cu interconnections was proposed by Sukharev and Zschech [21]. Instead of analyzing the vacancy flux in the EM Eq. 2.1, a mass balance equation for atomic concentration was adopted in their model as follows.

$$\frac{\partial N}{\partial t} + \nabla J = 0 \tag{2.21}$$

where $N(r, t) = N/N_0 = \Omega N$ is the normalized atomic concentration at the $r = (x, y, z)$ location at the moment of time t, $\Omega = 1/N_0$ is the atomic volume with N_0

as the initial atomic concentration, and J is the total atomic flux at this location. In their study, various driving forces were incorporated into the mass balance equation, such as the concentration gradients, the electron wind force, temperature gradients, and the mechanical stress gradient as follows.

Atomic flux due to concentration gradients was given as

$$J_N = -D\nabla N \tag{2.22}$$

Atomic flux due to electron wind force was given as

$$J_C = \frac{NeZ^*D\rho}{kT}j \tag{2.23}$$

Atomic flux due to temperature gradient was given as

$$J_T = -\frac{NDQ}{kT^2}\nabla T \tag{2.24}$$

Atomic flux due to mechanical stress gradient was given as

$$J_s = \frac{ND\Omega}{kT}\nabla \sigma_h \tag{2.25}$$

Combining Eq. 2.21 and Eqs. 2.22–2.25, the atomic concentration at any point of the considered segment can be characterized for a given current density, temperature and stress gradients. In order to get a complete solution of the model, they determined the evolution of the current, temperature, and stress distributions in the considered segment. Similar to the model developed by Gleixner and Nix [20], different diffusivity values were introduced to simulate the effect of microstructure and surrounding material on atomic flux based on the process technology. The diffusivities along different interfaces, grain boundaries, and inside the bulk were treated differently to simulate the changes in the failure mechanism due to different process technologies. The predicted locations of the void nucleation sites as well as the predicted void dynamics in a variety of interconnect segments fitted well to the available experimental data [22, 23].

2.1.2.2 2D Void Surface Evolution Model

The simulation methodology for this class of model was initiated in the mid-1990s, and two different modeling methodologies were developed. Both of them are 2D models based on finite element analysis to calculate the governing equation for current density, temperature, and stress distributions in a 2D manner. One is the *sharp interface model* that attempts to model the void surface as a sharp interface between the conductor material and empty space in the void. This method requires a tracking and accurate book-keeping of the surface elements and their geometry, and hence it is cumbersome. However, it is able to explain some experimental observations well.

The other method is the *phase field model*. This model defined the level of the presence of the conductor material based on a phase field scalar variable.

The value of the variable defines the phase of the material at any point on a fixed grid that defined the simulated specimen. The transition from metal material to an empty void is not sharp as modeled by the first approach, but gradual in terms of the presence of the material. It requires less book-keeping and seems to be more computationally efficient.

2.1.2.2.1 Sharp Interface Model

In 1994, Arzt et al. [24] published their work on the investigation of the behavior of EM-induced voids in narrow, unpassivated aluminum interconnects. Experimentally, it was found that the fatal voids have a specific asymmetric shape with respect to the electron flow direction. They proposed, to the best of our knowledge, the first model which attempted to simulate the void shape changes on the basis of atomic diffusion along the void surface. In their model, they considered an isolated 2D void which extends through the thickness of the line, and atomic diffusion on the void surface was assumed to be the primary transport mechanism. Depending on the balance of the arriving and departing atoms at each point on the surface, the movement and the shape change of a void, in the line was tracked as shown in Fig. 2.4.

The driving forces for such a behavior were modeled using the surface mass flux through a surface element; and they considered two types of mass fluxes as follows

$$I_s^j = -\frac{\delta D_s}{\Omega k T} e Z^* \rho j_s \tag{2.26}$$

$$I_s^\kappa = \frac{\delta D_s \gamma}{kT} \frac{\partial \kappa}{\partial s} \tag{2.27}$$

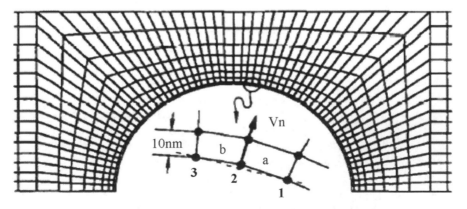

Fig. 2.4 The finite element analysis for a sharp interface model. Reprinted from Kraft and Arzt [25], copyright © 1997, with permission from Elsevier

where I_s^j and I_s^κ are the mass fluxes along the surface due to the electron wind force and the curvature of the surface, respectively, δ is the film thickness, D_s is the surface diffusivity, Ω is the atomic volume, k is the Boltzmann constant, T is the absolute temperature, e is the charge of an electron, Z^* is the effective valence, ρ is the resistivity, j_s is the current density along the surface, γ is the surface tension, κ is the curvature, and s is the arc length along the surface.

In their model, the current density distribution was calculated based on the assumption of a steady flow of an incompressible, non-viscous, circulation-free liquid, of which the expressions are mathematically identical to the equations of electrostatics. With the knowledge of current density and mass fluxes, the normal velocity at each point of the surface can be computed and the resulting equation of motion of the void surface can be obtained as

$$v_n = \Omega \frac{\partial \left(I_s^j + I_s^\kappa \right)}{\partial s}. \tag{2.28}$$

However, with their methodology, the inhomogeneous current distribution in the interconnect around the void was not considered in their model. This non-uniform current density distribution produces temperature gradients, and thus results in local changes in the resistivity and diffusivity. These effects are important for the electromigration failure mechanism in narrow interconnects.

The simulation methodology was thus further modified in their subsequent report [25]. A FEM was used in the calculation of the current density and the temperature distribution. A finite difference method was employed for the void motion and the shape changes computation. Figure 2.5 shows some of their simulation results of the void motion and the shape changes in an Al metal line.

Fig. 2.5 Simulation of the development of initially semi-circular voids with radius of **a** 0.8 μm, **b** 0.6 μm in a 1 μm wide Al metal line. Reprinted from Kraft and Arzt [25], copyright © 1997, with permission from Elsevier

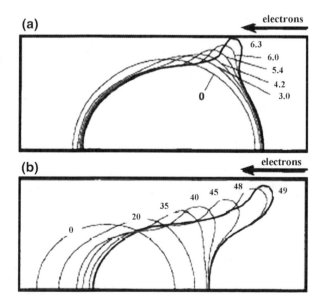

The commercial finite element software ANSYS was used in their modeling. Also included in their work was the consideration of the crystallographic orientation leading to a surface diffusivity change for different angles of the surface with respect to the crystal. To model this effect, they incorporated the anisotropy of the surface diffusivity in the finite element analysis by describing the surface diffusivity as a function of the angle θ between the surface tangent and the conductor line length direction.

In their study, they found that slit voids formed within the grains are likely to be caused by EM-induced surface diffusion. Under sufficiently high current density, a rounded void is unstable and will spontaneously collapse into a slit. The procedure used by Kraft et al. [25] is shown in Fig. 2.6 where an initial void shape is defined, followed by the computation of the current density and temperature due to Joule heating, then the calculation on the shape changes of the void surface. Thereafter, a re-meshing is needed for the modified structure, and the process repeats itself. With this procedure, void shape evolution and shape stability with different void geometries can be studied, and the model has the potential of predicting the lifetime of metal line with certain void shape.

Another similar model was proposed by Wang et al. [26]. In their early studies [27, 28], they showed that atomic diffusion on the void surface, driven by the electrical current, can cause a circular void to translate into a slit void. During this translation process, two forces compete in determining the void shape; one is the surface tension force and the other is the electron wind force due to the electrical current. Surface tension force favors the formation of a rounded void, while electrical current favors a slit void; a rounded void will collapse and become a slit when the electron wind force dominates.

In their later published work [26], they reviewed the experimental and theoretical findings, and provided a numerical simulation of the void shape change.

Fig. 2.6 The flowchart of numerical simulation of the void shape change. Reprinted from Kraft and Arzt [25], copyright © 1997, with permission from Elsevier

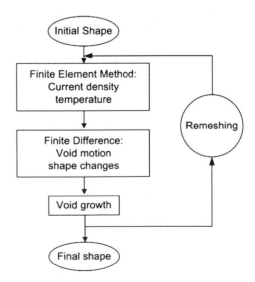

Similar to the Kraft's model, they adopted the sharp interface approach and assumed that the void shape change was due to the surface diffusion only, considering all other transport processes as negligibly slow during the void shape change. This approach utilized the same equations as in Kraft's model [24] to model the mass transport. By approximating the void perimeter by many short straight segments, they formulated a finite element procedure for the shape evolution, as done by Kraft [24].

The difference between the above-mentioned two models lies in the implementation of the physical system analysis. Wang et al. used a conformal mapping technique to determine the electric field around the circular void. After that, the electric potential and the curvature of the void surface were determined, and then the Galerkin method [29] was used to determine the EM kinetic system. The Galerkin method will be elaborated in Chap. 3. As a result, their model is capable of modeling void and surface evolutions, and their results are similar to that obtained by Kraft et al. [25]. Compared with the FEM by Kraft, Wang's approach, while mathematically elegant, is not versatile and is hard to yield more complex geometry evolutions [30]. Other similar works were reported by Gungor and Maroudas [31, 32] and Schimschak and Krug [33].

In the sharp interface models developed by many, the common assumption is that the atomic surface diffusion is the dominant diffusion mechanism, and the surface diffusion is driven by the electron wind force and the surface tension force due to the curvature of the void surface only. The anisotropy of the void surface diffusivity is emphasized, and slit voids will only form in grains with certain crystallographic orientations. They studied moving boundary problems entail explicit tracking of the boundary. The interface is described by specifying a large number of points on it. The same equations were used by them to model the atomic surface diffusion and the motion of the void surface at each interface point.

2.1.2.2.2 Phase Field Model

In the Sharp interface model, a lot of interface points are required to accurately describe the void surface when the void shape is evolving, and the changing boundary conditions must be implemented at this increasingly numerous set of points. Thus, the sharp interface model can get very complicated and also tends to have rather poor numerical stability [34, 35].

On the other hand, if the entire domain is described by a continuously varying scalar order parameter ϕ which has a value of $+1$ for region well within the metal "phase" and -1 for region well within the void "phase", and ϕ has a value between $+1$ and -1 for the metal–void interface as shown in Fig. 2.7, we have the phase field model [36].

This model was introduced independently by Fix [37] and Collins and Levine [38], and it received considerable attention in the context of phenomena associated with evolving interface. The earliest attempt to use the phase field model for the line

Fig. 2.7 The void simulation in a phase field model

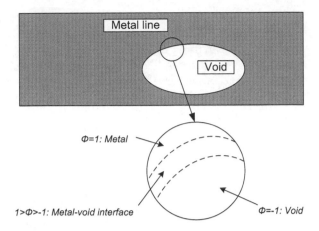

interconnect failure simulation was reported by Mahadevan and Bradley [35]. In their study, they simulated the time evolution of a perturbation to the edge of a current carrying, single-crystal, unpassivated metal line. Surface electron wind force migration, surface self-diffusion due to the current crowding, and the curvature of the void surface were all taken into account. They adopted the same formula that model the diffusion mechanisms along the void surface used by Kraft [24], but the idea of a sharp interface between metal and void was abandoned. The inclusion of the phase field model results in two coupled partial different equations, with the first one describing the dynamics of the phase field, and the second one describing the electric field as given below.

$$-\varepsilon\frac{\partial\phi}{\partial t} = \vec{\nabla}\cdot\left\{\alpha_0\left[1 + \lambda_s\varepsilon\left(\vec{\nabla}\phi\right)^2\right]\vec{\nabla}v - \left[\varepsilon^2\beta_0(\phi+1)\left(\vec{\nabla}\phi\right)^2\vec{\nabla}\Phi_E\right]\right\} \quad (2.29)$$

$$\alpha_0 = \frac{3M_0\lambda_s\Omega}{D_sL}, \quad \lambda_s = \frac{3D_sh}{M_0Lk_BT}, \quad \beta_0 = \frac{3Z_s^*eE_0h}{k_BT} \quad (2.30)$$

$$v = \varepsilon^2\nabla^2\phi + g(\phi), \quad g(\phi) = (\phi - \phi^3)/2 \quad (2.31)$$

and

$$\vec{\nabla}\cdot\left[\sigma(\phi)\vec{\nabla}\Phi_E\right] = 0 \quad (2.32)$$

where D_s is the surface diffusivity and L is a characteristic length describing the void which can be taken to be the square root of the void surface area. Φ_E is the electrical potential, ε is a dimensionless constant proportional to the interface width, M_0 is the lattice diffusivity of the phase field, k_B is the Boltzmann constant, λ_s is the surface tension, T is the temperature, and σ is the conductivity.

The phase field model was used for an isolated void in an infinite thin film, and this method can be easily extended so that it applies to the time evolution of an edge perturbation in a metal line of finite extent. By solving the phase field equations and the electric field equations numerically, the model provides the time evolution of a small notch at the edge of a current-carrying single crystal metal line [35]. The model is able to predict a threshold value of the applied current so that the edge perturbation will grow into a slit-shaped void that spans the wire. Mahadevan and Bradley also explained the physical origin of this instability and pointed out the importance of the crystalline anisotropy and mass transport along the edge of the line.

At nearly the same time as Mahadevan and Bradley, Bhate et al. [39] reported their own version of the phase field model for simulating the process of electron wind force migration, curvature-driven surface migration, and the stress-driven migration. In their work, Bhate et al. [39] briefly discussed the theory of the sharp interface model and its limitations and disadvantages. They then proposed their own phase field model. Their approach is based on the introduction of an order parameter field to characterize the damaged state of an interconnect. The order parameter takes on distinct uniform values within the material and the void, varying rapidly from one to the other over narrow interfacial layers associated with the void surface. They derived the field equations for the order parameters based on the microforce balance principle of Gurtin [40].

Accordingly, they expressed the free energy of the line as

$$
\mathbb{F}(\phi, \varphi) = \int_R F(\phi, \nabla\phi, \varphi)dV = \int_R \left\{ \frac{2\gamma_s}{\varphi\pi}\left(f(\phi) + \frac{1}{2}\varphi^2|\nabla\phi|^2\right) + W(\varphi, \phi)\right\}dV
$$

(2.33)

where $f(\phi)$ is the bulk free energy, γ_s is the isotropic surface energy, φ is a parameter that controls the thickness of the interfacial layer associated with the void surfaces, and W is the elastic strain energy. The atomic diffusion paths through the bulk and grain boundary were assumed to be negligible, and the only mode of transport for atoms was the diffusion along the void surfaces in their model. This is the same assumption as in most sharp interface models, except that the surface diffusivity in their model was treated as isotropic. However, this assumption is not justifiable since it has been proved that the anisotropic diffusivity of atoms along the void surface has a strong effect on the evolution of void during EM [24].

2.1.3 Summary

From the above discussion, all the 1D EM models are focused on the simulation of the EM kinetics. The kinetics of EM, such as the EM driving forces due to electron wind force and vacancy concentration and stress gradients, were simulated through the calculation of the change in vacancy concentration along the metal line. From the evolution of the vacancy concentration, the critical location of the metal line can be identified, and the time to failure can be extracted if reasonable criterion of failure is given.

The 1D EM modeling is capable of integrating the multiple driving forces to simulate the kinetic of EM process and evaluating the impact of the properties of the metal line on the EM performance, such as the length, the passivation layer, and the various boundary conditions that represent the interconnect structures in a simplified manner. However, the real EM process in a metal line is rather complicated. 1D EM modeling is only capable of simulating the simplified scenarios of EM due to its limitation on dimensionality. The influence of the triple points of the multiple grain textures, the surrounding materials and the induced thermo-mechanical stress, the shape of the metal line or the microstructure of the interconnects were poorly addressed in the 1D modeling. The consideration of the EM from thermodynamics aspect, such as the various diffusion paths for atoms diffusion, was highly simplified in the 1D model.

Unlike the 1D EM models, 2D EM models are able to consider the effects of microstructure of interconnects and the surrounding material on the EM performance. However, the difficulties and complexities in the mathematical implementation increase dramatically as the coupled partial differential equations have to be solved in multiple dimensions. Rather than tracking the vacancy concentration evolution of a 1D metal line, most of the 2D EM models are based on the formulation and calculation of the atomic flux due to various driving forces and along various diffusion paths in the metal line. The typical EM-driving forces include electron wind force, atomic concentration gradient, temperature gradient, stress gradient and surface tension. The typical EM diffusion paths include interface diffusion, grain boundary diffusion, and lattice diffusion inside a grain. Numerical method, such as finite element analysis, is needed to solve the EM equations in a coupled manner with current density, temperature, and stress distributions.

Most of the proposed 1D and 2D models investigate the EM within the interconnect material itself, and the effect of the surrounding materials on the interconnect EM is not taken into consideration. However, as the interconnect line width goes into 150 nm and below, and with the use of the low-k dielectric, the impact of the surrounding materials on the interconnect EM becomes significant. The thermal conduction and the mechanical stiffness of the dielectric materials surrounding the interconnect structure play a major role in determining the EM reliability of an interconnect system. In other words, EM studies can no longer be limited to the interconnect material itself, and one needs to consider the entire interconnect system. Such a study will require a 3D EM simulation. A more detailed discussion on the application of finite element analysis for EM simulation in 3D manner will be given in Chaps. 3 and 4 of this book.

2.2 Review on the Modeling of the Stress-Induced Voiding (SIV)

Stress-induced voiding is another serious reliability challenge to IC interconnects for both the Al- or Cu- based metallization. Large thermo-mechanical stress is developed in interconnects during fabrication and cooling to room and operating temperatures, and this initiates stress relaxation and give rise to SIV. The physical mechanism of the stress development and relaxation have been studied [41, 42]. For example, chip with electroplated Cu interconnects consists of small grains initially. If the lines are not annealed properly before encapsulation, subsequent annealing will cause grain growth, resulting in grain growth-induced stress in the line [43]. The stress-induced voids can deteriorate the EM lifetime of an interconnect as observed experimentally [44].

The failure site varies in interconnects under thermo-mechanical stress. Ogawa et al. reported SIV void formation under via due to grain growth without thorough annealing [45]. The void in via in the form of copper pull up is attributed to poor interface adhesion in Cu/FSG interconnects [46]. Hommel et al. also reported SIV voids along via side walls and via bottom for Cu interconnects [47]. In the experimental work by An and Ferreira [48], dislocation was observed in wide interconnect while SIV void was formed in narrow interconnects. Via deformation was also observed in Cu/SiLK line-via structure [49]. In fact, the type and location of the failure is strongly dependent on the interconnect geometry, dielectric type, as well as the microstructure of the interconnects.

Conventional stress measurement methods such as wafer curvature measurements [50, 51], analytical modeling [12], and X-ray diffraction (XRD) methods have been developed to analyze the stress distribution in interconnect systems [52]. However, these methods are limited to simple test structures and they can only determine the average stress in the interconnect. On the other hand, FEM can be performed to determine the stress condition of complex multi-level structures [53, 54]. In this section, we will present a comprehensive review on the physics-based SIV modeling in the past two decades. The review focuses on four categories: (1) thermo-mechanical stress modeling; (2) analytical modeling; (3) vacancy and atomic migration modeling; and (4) stress-induced migration modeling. Each category provides a unique perspective to help us understand the SIV phenomenon in interconnects.

2.2.1 Thermo-mechanical Stress Modeling

The characteristics of thermo-mechanical stress in interconnects depend on how effective the Si substrate and the dielectrics confine the metal interconnect from expanding freely. Therefore, the thermo-mechanical properties of the dielectrics and the geometry of the structure can affect the magnitude and nature of the thermo-mechanical stress greatly. Compared with SiO_2, low-k materials are expected to have poorer mechanical properties and this can lead to severe stress-related problems in low-k interconnects [55].

Hydrostatic stress and von Mises stress are two important quantities in studying the stress-induced failure in interconnects. The hydrostatic stress σ_H and von Mises stress σ_V are given as [56]

$$\sigma_H = \frac{\sigma_1 + \sigma_2 + \sigma_3}{3} \tag{2.34}$$

$$\sigma_V = \frac{1}{\sqrt{2}}\left[(\sigma_1 - \sigma_2)^2 + (\sigma_2 - \sigma_3)^2 + (\sigma_3 - \sigma_1)^2\right]^{1/2} \tag{2.35}$$

where σ_1, σ_2, and σ_3 are the principal stresses in the line along the stress orthogonal caterisan coordinates.

Hydrostatic stress is proven to be the driving force for void nucleation [57]. On the other hand, von Mises stress, which is normally used as a criterion for evaluating deformation, does not result in the volumetric changes of the material [58]. At the early stage of a typical SIV process, under the constraint of the surrounding materials, the shear stress in an interconnect is relaxed through either diffusion creep or dislocation glide [59, 60], causing the interconnect to be at a near perfect hydrostatic state. During the subsequent void-growing process, the duration of plastic deformation is found to be much shorter than the duration of the hydrostatic stress relaxation [59], and von Mises stress can therefore be neglected. In fact, a gradient in hydrostatic stress can still exist after plastic deformation, which induces SIV subsequently [61].

In the work of Rhee et al., stress behavior of blanket Cu films were studied through both X-ray diffraction method and FEM [53]. With the help of the FEM, the effects of scaling, barrier thickness, and low-k dielectrics were studied in their work. It was found that the hydrostatic stress normal to the surface increase with decreasing line width while the stress along the line did not change much with line width. Besides, the severity of SIV formation in low-k interconnect is reduced due to lesser confinement effect by the surrounding dielectrics. This is due to the fact that the effective bulk modulus is much smaller for low-k interconnects than that of SiO_2-based interconnects as is also reported in the work of Hou and Tan [62] using

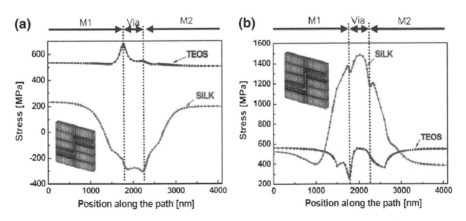

Fig. 2.8 Stress variation from M1 to M2 along the path for the TEOS-embedded and the SiLK-embedded structures. **a** Hydrostatic stress, **b** von Mises stress. Reprinted from Paik et al. [56], copyright © 2004, with permission from Elsevier

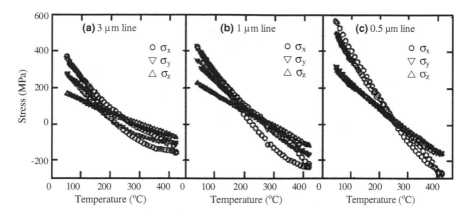

Fig. 2.9 Stress versus temperature curve of passivated Al(Cu) lines for **a** 3 μm wide, **b** 1 μm wide, and **c** 0.5 μm wide lines, respectively [59], copyright © 1995 IEEE

FEM. On the other hand, the increases in the von Mises stress in low-k interconnect raises the reliability concern for interfacial delamination or plastic deformation in low-k passivated lines [53].

The stress state in M1-via-M2 structures is studied by Paik et al. [56] using FEM. Two extreme cases were considered in their work, namely TEOS- and SiLK as dielectrics respectively. The hydrostatic stress and von Mises stress distributions along M1, via and M2 for TEOS and SiLK-based interconnects are shown in Fig. 2.8a, b respectively. It was noted that the von Mises stress of the SiLK structure was much higher than that of the TEOS structures in the via, although the von Mises stresses in the M1 and M2 lines of SiLK structure were similar to those of TEOS structures as shown in Fig. 2.8b. This implies that the stress state of via is significantly different from that of lines and deformation will occur mainly in the via and not in the lines. Similar modeling work can also be found in other references [55, 63].

The scaling effect on the thermo-mechanical stress in interconnect was addressed in the work of Ho et al. through beam bending measurement [59]. It was found that with decreasing line dimensions, the confinement due to the surrounding materials was enhanced and the stress level was increased sufficiently to cause void formation. In their bending beam measurement experiment, Al line with line width of 3, 1 and 0.5 μm were subjected to thermal cycling from room temperature to 400°C. The lines behaved more elastic with less stress hysteresis during thermal cycling with smaller line width as shown in Fig. 2.9. With decreasing line width, the difference in the magnitude of the stress components decreases, making the stresses more hydrostatic. For the stress relaxation behavior, the rate of stress relaxation along the line was the fastest, followed by the stress component across the line and then the stress normal to the line.

Compared with FEM, analytical modeling of stress components is more useful to extract quick estimates and to gain a greater insight into the mechanics and physics of the stress evolution. The early analytical models were based on the

Eshelby theory of inclusions in which the line was modeled as an ellipsoidal cylinder embedded in an infinite isotropic matrix of the passivation material [60, 64]. However, these models did not capture the interaction between neighboring lines as well as the elastic properties of the substrate or sharp edges of the line. These two limitations were later overcome by Wikstrom et al. [65] and the analytical modeling results were found to be consistent with the FEM calculations.

2.2.2 Analytical Modeling of SIV

Pre-existing micro-voids were found in both wide and narrow Cu interconnects under thermo-mechanical stress before electromigration (EM) test [66]. These microvoids can grow during EM and become fatal voids at the cathode end of an interconnect [44]. Shen et al. found that a large stress-induced void was more prone to growth during subsequent EM test [67]. Many studies have been carried out to investigate SIV from different perspectives such as the quality of passivation layer [68], the mechanical strength of interlayer dielectrics (ILD) [53], the microstructure of the film [69], and they were indeed found to affect the thermo-mechanical stress in metallization and hence its SIV lifetime. Detailed information of such factors is incompletely known however, and will probably never be precisely controlled in fabrication. There is no single model that accounts for all the aspects of SIV. Therefore, it is desirable to consider the physics of the phenomenon semi-quantitatively and different analytical models will be discussed here.

2.2.2.1 Saturate Void Volume

An encapsulated metal line is stress free at high temperature T_0, and it is commonly known as stress free temperature (SFT). As the total number of atoms in the line is constant at all time, when the line is cooled down to storage temperature T, the thermo-mechanical stress is generated due to the mismatch in the coefficient of thermal expansion (CTE) of the interconnect and the surrounding materials. Void space is then created in order to release the resulting stress, and upon complete relaxation, the void volume will be saturated. The saturated void volume (V_{SV}) is given by Suo [70]

$$V_{SV} = 3 \cdot \Delta\alpha \cdot (T_0 - T)V \tag{2.36}$$

where V is the volume of the metal line and $\Delta\alpha$ is the effective thermal expansion mismatch strain, i.e., the difference in the CTE between the metal line and its surrounding materials. For silica-based interconnect structure, since the metals have much larger CTE than that of the surrounding materials, we can use the CTE of Cu in the estimation. Using the typical values of 18×10^{-6}/K for $\Delta\alpha$ and 300 K for $(T_0 - T)$ [70], the void volume fraction is evaluated to be 1.62% for Cu interconnects.

2.2.2.2 Void Initiation

Within the context of continuum mechanics, SIV is one of the means to release the stored strain energy of an interconnect system by mass redistribution. Considering a tiny void in a metal line under thermo-mechanical stress, the void can change its size by relocating atoms on its surface to area in the metal far away from the void under stress gradient-induced driving force. In this process, the void increases its surface area and hence the free energy increases. During this process, work is done by the thermo-mechanical stress, reducing the free energy of the system.

Considering interconnects without tiny void under the thermo-mechanical stress as the ground state, the system has zero free energy. Now, the current state has a void of radius a. Assuming the void is in spherical shape, and let γ be the surface energy per unit area. By creating the void, a surface of area of $4\pi a^2$ is exposed, raising the free energy of the system by $\gamma 4\pi a^2$. At the same time, atoms occupying the volume of $4\pi a^3/3$ is relocated from the void, allowing the remote stress to do the work by $\sigma 4\pi a^3/3$. Thus, the free energy of the system, relative to that of solid with no void is

$$F(a) = 4\pi a^2 \gamma - \frac{4}{3}\pi a^3 \sigma \qquad (2.37)$$

Figure 2.10 shows the free energy as a function of the void radius. Based on Eq. 2.37, the critical void radius is given by a* $= 2\gamma/\sigma$. One can see from Fig. 2.10 that when the void is small, the surface energy dominates, and the void will shrink to reduce the free energy, rendering no void formation after some time. When the void is large, the stress dominates, and the void will grow to reduce the free energy. The critical void radius a* exists when the free energy is maximum. Using typical values of 1 J/m^2 for γ and 200 MPa for σ in Al(Cu) material [71], the critical radius is evaluated to be 10 nm. Such void size is used in the work of Hou and Tan for the simulation of thermo-mechanical stress in the presence of the mini-void [71].

Fig. 2.10 The free energy as a function of radius. Reprinted from Suo [70], copyright © 2003, with permission from Elsevier

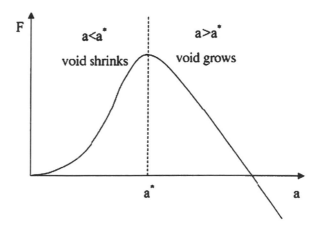

2.2.2.3 Lifetime Formulations

The stress evolution during SIV can be described by the following [72, 73],

$$\frac{\delta \sigma_H}{\delta t} = K \nabla^2 \sigma_H \tag{2.38}$$

where t is the time, σ_H is the local hydrostatic stress driving the mass transport, and K is the effective diffusivity defined as $K = \frac{DB\Omega}{k_B T}$. Here, D is the atomic diffusivity, Ω is the atomic volume, k_B is the Boltzmann's constant, T is the local temperature, and B is the effective bulk modulus which describes the confinement effect of the surrounding materials on the metal line. Equation 2.38 governs the stress evolution in interconnects. The analytic solutions of Eq. 2.38 for 1D and 2D line structures can be found in the works from Zhai and Blish [73]. For the case of 1D, it is given as

$$\sigma(x, t) = \frac{2\sigma_0}{m_n} \sum_{n=0}^{\infty} \exp\left(-K m_n^2 t / L^2\right) \sin\left(\frac{m_n}{L}x\right) \tag{2.39}$$

Equation 2.39 represents the analytical stress solution for 1D line structure where $m_n = (2n + 1)\pi/2$ and L is the length of the interconnect. This 1D model represents the intermediate and later stage for a typical SIV process where the diffusion length is much larger than the line width.

For a 2D line structure, the solution of Eq. 2.38 is given as

$$\sigma(r, t) = \sigma_0 \cdot erf\left(r/\sqrt{4Kt}\right) \tag{2.40}$$

Equation 2.40 is the analytical stress solution for 2D line structure where $r = 0$ is the origin or the void nucleation site. This 2D model represents the early stage of SIV where the diffusion length is shorter than the line width.

Equations 2.39 and 2.40 form a simplified stress evolution model for SIV, which contains some of the most fundamental perspectives including effective diffusivity, line length, time, and initial stress.

Assume an initial flaw exist in a metal line and is large enough that the stress near the flaw is zero. In this case, Eq. 2.38 is identical to the usual diffusion equation with K acts as the diffusivity. The solution to this initial-boundary value problem is well known, and the time for a void to relax a segment of interconnect of length l is [70]

$$t_l \sim \frac{L^2 kT}{DB\Omega}. \tag{2.41}$$

Assume a void length of L constitutes the failure, the SIV lifetime is approximated as [70]

$$t_{life} \propto \frac{1}{B[(\alpha_m - \alpha_d)(T_0 - T)]^2} \exp\left(\frac{E_A}{k_B T}\right). \tag{2.42}$$

This effective bulk modulus B is dependent on the cross section of the interconnect as well as dielectric materials, and one can therefore inferred from Eq. 2.40 that a structure with a smaller B will lead to a longer SIV lifetime, if other things being the same.

Fischer et al. proposed another model to formulate the SIV lifetime [47, 74]. Their derivation was based on the assumption that the change in the plastic strain leads to an increase of the void volume which is proportional to the increase of the line resistance. With this assumption, they derived the median time to failure (MTF) for SIV as follows

$$t_{life} \propto C \frac{T}{T_0 - T} \exp\left(\frac{E_A}{kT}\right) \tag{2.43}$$

where C is independent on temperature and represents a specific constant for a set of identical test structures.

The temperature exponents in Eqs. 2.42 and 2.43 are different, and different temperature exponent can lead to a deviation of the estimated activation energy. In the recent work by Hou et al. [71], the SIV lifetime formulations were derived from the energy perspective. They found that the temperature exponent in the SIV lifetime formulations was determined by the available diffusion paths for the interconnect atoms and the interconnect geometry. Based on the diffusion pathways available, they classified the SIV process into three categories as linear, square, and cubic as shown in Fig. 2.11a–c.

The generalized expression for the SIV lifetime is expressed as [75]

$$t_{life} \propto C \frac{T}{(T_0 - T)^N} \exp\left(\frac{E_A}{k_B T}\right) \tag{2.44}$$

where N is the temperature exponent depending on interconnect geometry and microstructures and C is independent on temperature. N is equal to 4, 2, and 4/3 for the linear, square and cubic cases respectively.

Equation 2.44 implies that the failure time approaches infinity when SIV test is carried out at either the SFT T_0, or the absolute 0 K. As the test temperature increases from 0 K, the tensile thermo-mechanical stress decreases linearly and the

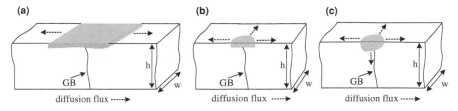

Fig. 2.11 Three cases of stress relaxation volume evolution are indicated by the *grey* area. The relationship between the stress relaxation volume and diffusion length is categorized as **a** *linear*, **b** *square*, and **c** *cubic*. Reprinted with permission from Tan and Hou [75], copyright © 2007, American Institute of Physics

diffusivity increases exponentially. As a consequence, the SIV lifetime decreases with temperature up to a critical point, above which the lifetime increases with increasing temperature. Therefore, to maximize acceleration in reliability test, SIV should be performed around T_{crit}. This T_{crit} can be determined by differentiating Eq. 2.44 with respect to T, and we have

$$T_{crit} = \frac{2E_A T_0}{\sqrt{(E_A - k_B T_0)^2 + 4E_A T_0 k_B N} + E_A + k_B T_0}. \tag{2.45}$$

Equation 2.45 reveals that T_{crit} is dependent on the activation energy, temperature exponent, and SFT.

Compared with the work of Ogawa et al. [76], where E_A was given as 0.74 eV and T_0 was found to be 270°C, T_{crit} is evaluated to be within 172–228°C using Eq. 2.45 for the different value of N used, and the computed T_{crit} agrees well with their experimental observations of 190°C. In the recent work of An et al. [48] where the grain boundary diffusion was found to be as important as the interface diffusion, we use the temperature exponent of 4/3 for Eq. 2.45 with E_A at 0.74 eV as in the previous case, the critical temperature evaluated from Eq. 2.45 is around 255°C, in good agreement with the experimentally observed 250°C. For Al interconnects, Fischer et al. also reported the similar critical temperature for Al interconnects at slightly higher activation energy [74].

2.2.3 Vacancy and Atomic Migration Model

SIV phenomenon occurs during storage test for long periods and is accelerated at high temperature storage. Numerous vacancies are generated thermodynamically in the metal interconnect at high temperatures, and they migrate due to the stress gradient and accumulate at vacancy sinks in the interconnect. The accumulation of the vacancies leads to the formation of voids. Based on the fact that vacancy migration and void growth are enhanced by the thermo-mechanical stress, stress-induced vacancy model was proposed in Aoyagi's early work [77]. In his work, the vacancy distribution was directly correlated with the thermo-mechanical stress distributions based on the stress-induced vacancy model, and the thermo-mechanical stress distribution was simulated using FEM. The SIV weak point was predicted to be near interconnect sidewalls and it was in qualitative agreement with the experimental results for Al interconnects.

The temperature characteristics of SIV was studied through the model of Aoyagi [78]. The vacancy concentration C under stress is given by Flinn [57],

$$C = \exp\left(-\frac{E_F}{k_B T}\right) \exp\left(\frac{S_F}{k_B}\right) \exp\left(\frac{\sigma \Omega}{k_B T}\right) \tag{2.46}$$

where E_F is the single-vacancy formation energy, S_F is the single-vacancy formation entropy, Ω is the vacancy volume, and σ is the stress which is defined to be positive for tensile stress. Here, the vacancy concentration C is defined as the ratio of the number of vacancies to the number of atoms constituting the metal.

When vacancy sinks (e.g. defects) exist in the interconnect under tensile stress, the vacancies migrate to the vacancy sinks. The driving force of the migration of vacancies is the difference of the stresses on the void surface which is zero and the residual thermo-mechanical stress in the interconnect. Aoyagi's study revealed that the temperature characteristics of the vacancy flux has a peak at a certain temperature which changes due to stress relaxation [78].

Later on, Aoyagi also performed numerical simulation to evaluate the resistance change in interconnect during the SIV process [79]. He considered a simplified slit-like void as shown in Fig. 2.12. The atomic migration model was based on the migration of atoms due to the thermo-mechanical stress gradient present in interconnects.

The atomic flux J in the SIV process can be expressed as [79]

$$J = \frac{D}{k_B T} N_{atom} \nabla (\sigma \Omega) \qquad (2.47)$$

where D is the atomic diffusion coefficient, N_{atom} is the number of atoms, Ω is the atomic volume, and σ is the thermo-mechanical stress which is defined to be positive for a tensile stress. In general, the diffusion coefficient is enhanced at a higher temperature or under a tensile stress as given by Aoyagi [79]

$$D = D_0 \exp \left(-\frac{E_m}{k_B T} \right) \exp \left(\frac{\sigma \Omega}{k_B T} \right) \qquad (2.48)$$

where E_m is the activation energy for the self-diffusion of atoms and D_0 is the diffusion constant. The relationship between resistance change, storage temperature, storage time, and void width is shown in Fig. 2.13a–d. We can see that the interconnects disconnect suddenly after a certain storage time, depending on the storage temperature.

Fig. 2.12 *Top view* of an interconnection with a slit-like void for the simulation. Reprinted from Aoyagi [79], copyright © 2006, with permission from Elsevier

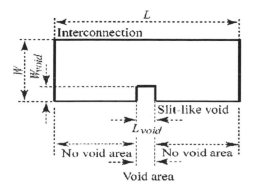

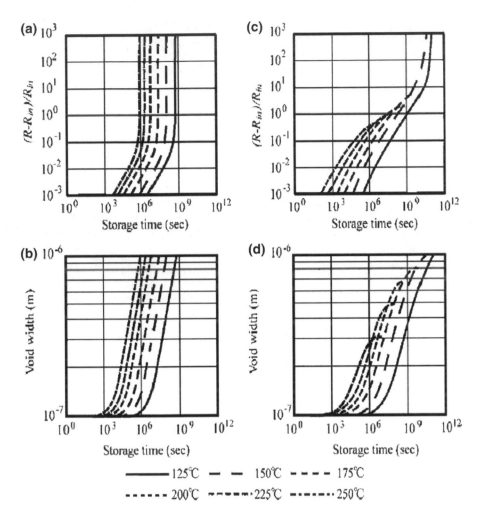

Fig. 2.13 Change in the line resistance and void width as a function of storage time for aluminum interconnect with the storage temperature as a parameter. **a, b** The void is 1 nm long and 100 nm wide in its initial stage. **c, d** The void is 50 nm long and 100 nm wide in its initial stage. Reprinted from Aoyagi [79], copyright © 2006, with permission from Elsevier

2.2.4 SIV for Nano-interconnects

Replacing SiO_2 with low-k materials results in a decrease of the tri-axial stress in Cu [80]. However, it is still under debate whether the SIV performance for low-k interconnects improves or degrades when compared with that in SiO_2-based interconnects. Bruynseraede et al. [81] stated that SIV shall improve due to a lower density of nano-defects generated by the plastic deformation in low-k interconnects. Suo [82] also reported that low-k interconnect should be more SIV resistant due to a reduced effective bulk modulus. However, Gan et al. [83] demonstrated that low-k interconnects suffered more from SIV at via bottom due to a higher stress gradient despite a lower thermo-mechanical stress.

Based on the recent SIV model proposed by Tan and Hou [75], SIV performance for the low-k interconnects depends on the compromise between the confinement and the passivation effects. Since the passivation condition is highly process dependent, if one can assume that the passivation on Cu surface is the same for both Cu/SiO$_2$ and Cu/low-k interconnects, Cu/low-k interconnects will outperform its counterpart in SIV due to a lower thermo-mechanical stress. In reality, the passivation condition is poorer for Cu/low-k interconnect as indicated by a lower interface energy for low-k interconnects [84]. In short, the SIV performance for Cu/low-k interconnects is strongly dependent on the process and materials used during fabrication.

When interconnect is scaling down to 0.16 μm, the thermo-mechanical stress will increase slightly as shown by Tan et al. [85] using FEM. Their simulation model is based on the classical continuum mechanics which is intrinsically size independent. However, when the line width is scaled down further into nano-regime, elastic properties of the interconnects become size dependent, and a qualitative departure from classical mechanics is expected. Sharma et al. [86] demonstrated that the thermo-mechanical stress decreases at line width goes below 50 nm by considering the size dependency in elasticity. The finite element analysis incorporating the size-dependent effects are still lacking for interconnects below 100 nm.

Besides the SIV, large von Mises stress in interconnects may also cause reliability problems, particularly for Cu/low-k interconnects. Paik et al. [63] showed that the von Mises stress in low-k interconnects is larger than that of the SiO$_2$-based interconnect, through finite element analysis. A large von Mises stress within the interconnect system would result in failures in the form of plastic deformation. For instance, via barrier layer cracking is reported by Paik et al. [56]

Fig. 2.14 Stabilized stress–temperature response for passivated Cu films of various thicknesses. Reprinted from Shen and Ramamurty [90], copyright © 2003, with permission from Elsevier

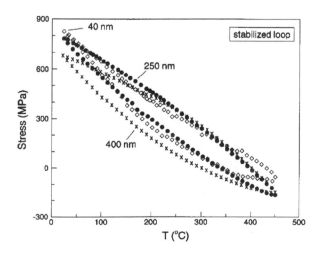

in low-k interconnect due to the high von Mises stress at via region. In fact, owing to the reduced fracture energy of the low-k materials [87], dielectric failure is becoming important for nano-interconnects with low-k dielectric [88].

As line width is scaling down, plastic deformation is less likely to happen due probably to the increased elasticity of the interconnects [89]. The enhancement in elasticity of interconnect was confirmed by Shen and Ramamurty [90] who found that the hysteresis loop becomes smaller with decreasing film thickness during thermal cycling as shown in Fig. 2.14, implying increasing difficulty of plastic deformation. Recently, Budiman et al. [52] also reported that the plastic deformation was less likely to happen in narrower interconnects by synchrotron X-ray microdiffraction measurements.

2.2.5 Summary

Stress-induced voiding is a serious reliability problem in metal interconnects. In this chapter, we review the progress on the physics-based SIV modeling from four aspects, namely the thermo-mechanical stress modeling, analytical modeling, vacancy and atomic migration model, and stress migration for nano-interconnects. Each category provides a unique perspective for us to understand the physics of SIV process.

References

1. Clement JJ (2001) Electromigration modeling for integrated circuit interconnect reliability analysis. IEEE Trans Device Mater Reliab 1:33–42
2. Rosenberg R, Ohring M (1971) Void formation and growth during electromigration in thin films. J Appl Phys 42:5671
3. Blech IA (1976) Electromigration in thin aluminum films on titanium nitride. J Appl Phys 47:1203–1208
4. Blech IA, Herring C (1976) Stress generation by electromigration. Appl Phys Lett 29:131–133
5. Shatzkes M, Lloyd JR (1986) A model for conductor failure considering diffusion concurrently with electromigration resulting in a current exponent of 2. J Appl Phys 59:3890
6. Kirchheim R, Kaeber U (1991) Atomistic and computer modeling of metallization failure of integrated circuit by electromigration. J Appl Phys 70:172
7. Clement JJ, Lloyd JR (1991) Numerical investigations of the electromigration boundary value problem. J Appl Phys 71:1729
8. Tu KN (1992) Electromigration in stressed thin films. Phys Rev B 45:1409
9. Ross CA (1991) Stress and electromigration in thin film metallization. Mater Res Soc Proc 225:35–46
10. Kirchheim R (1992) Stress and electromigration in Al-lines of the integrated circuits. Acta Metall Mater 40:309
11. Korhonen MA, Borgesen P, Tu KN, Li C-Y (1993) Stress evolution due to electromigration in confined metal lines. J Appl Phys 73:3790

12. Clement JJ, Thompson CV (1995) Modeling electromigration-induced stress evolution in confined metal lines. J Appl Phys 78:900
13. Park YJ, Andleigh VK, Thompson CV (1999) Simulations of stress evolution and the current density scaling of electromigration-induced failure times in pure and alloyed interconnects. J Appl Phys 85:3546
14. Park YJ, Thompson CV (1997) The effects of the stress dependence of atomic diffusivity on stress evolution due to electromigration. J Appl Phys 82:4277
15. Andleigh YK, Fayad W, Verminski M, Thompson CV (eds) http://nirvana.mit.edu/emsim/
16. Hau-Riege CS, Thompson CV (2000) The effects of microstructural transitions at width transitions on interconnect reliability. J Appl Phys 87:8467
17. Hau-Riege SP, Thompson CV (2000) Electromigration saturation in a simple interconnect tree. J Appl Phys 88:2382
18. Hau-Riege SP (2002) Probabilistic immortality of Cu damascene interconnects. J Appl Phys 91:2014
19. Tan CM (2010) Electromigration in ULSI interconnection. World Scientific Publishing Co, Singapore
20. Gleixner RJ, Nix WD (1999) A physically based model of electromigration and stress-induced void formation in microelectronic interconnects. J Appl Phys 86:1932
21. Sukharev V, Zschech E (2004) A model for electromigration-induced degradation mechanisms in dual-inlaid copper interconnects: effect of interface bonding strength. J Appl Phys 96:6337
22. Zschech E, Sukharev V (2005) Microstructure effect on EM-induced copper interconnect degradation: experiment and simulation. Microelectron Eng 82:629
23. Zschech E, Meyer MA, Mhaisalkar SG, Vairagar AV, Krishnamoorthy A, Engelmann HJ, Sukharev V (2006) Effect of interface modification on EM-induced degradation mechanisms in copper interconnects. Thin Solid Film 504:279
24. Arzt E, Kraft O, Nix WD, Sanchez JJE (1994) Electromigration failure by shape change of voids in bamboo lines. J Appl Phys 76:1563
25. Kraft O, Arzt E (1997) Electromigration mechanisms in conductor lines: void shape changes and slit-like failure. Acta Mater 45:1599
26. Wang W, Suo Z, Hao TH (1996) A simulation of electromigration-induced transgranular slits. J Appl Phys 79:2394
27. Suo Z, Wang W, Yang M (1994) Electromigration instability: transgranular slits in interconnects. Appl Phys Lett 64:1944
28. Yang W, Wang W, Suo Z (1994) Cavity and dislocation instability due to electric current. J Mech Phys Solids 42:897
29. Itô KE (1980) Methods other than difference methods. In: Iyanaga S, Kuwada Y (eds) Encyclopedic dictionary of mathematics, vol 2. MIT Press, Cambridge
30. Atkinson RR (2003) PhD, Rutgers, Multiphysics of degradation and failure of line interconnects. The State University of New Jersey, New Brunswick
31. Gungor MR, Maroudas D (1998) Electromigration-induced failure of metallic thin films due to transgranular void propagation. Appl Phys Lett 72:3452
32. Gungor MR, Maroudas D (1999) Theoretical analysis of electromigration-induced failure of metallic thin films due to transgranular void propagation. J Appl Phys 85:2233
33. Schimschak M, Krug J (2000) Electromigration-driven shape evolution of two-dimensional voids. J Appl Phys 87:695
34. Karma A, Rappel W (1996) Phase-field method for computationally efficient modeling of solidification with arbitrary interface kinetics. Phys Rev E 53:R3017
35. Mahadevan M, Bradley RM (1999) Simulations and theory of electromigration-induced slit formation in unpassivated single-crystal metal lines. Phys Rev B 59:11037
36. Mahadevan M, Bradley RM (1999) Simulations and theory of electromigration-induced slit formation in unpassivated single-crystal-metal lines. Phys Rev B 59:11037
37. Fix G (1983) Free boundary problems. In: Fasano A, Primicero M (eds) Research notes in mathematics, vol 2. Pitman, New York

38. Collins JB, Levine H (1985) Diffuse interface model of diffusion-limited crystal growth. Phys Rev B 31:6119
39. Bhate DN, Kumar A, Bower AF (2000) Diffuse interface model for electromigration and stress voiding. J Appl Phys 87:1712
40. Gurtin M (1996) Generalized Ginzburg-Landau and Cahn-Hilliard equations based on a microforce balance. Physica D 92:178
41. Sauter Mack A, Flinn P (1995) Effect of intrinsic passivation stress on stress in encapsulated interconnect lines. In: Materials Research Society Symposium, Boston, MA, USA, pp 465–470
42. Yeo I-S, Ho PS (1996) Stress relaxation and microstructural change in passivated Al(Cu) lines. In: 34th Annual proceedings of reliability physics symposium, Dallas, pp 131–138
43. Young-Chang J, Jong-Min P, Il-Mok P (2006) Effect of grain growth stress and stress gradient on stress-induced voiding in damascene Cu/low-k interconnects for ULSI. Thin Solid Films 504:284–287
44. Li C-Y, Borgesen P, Sullivan TD (1991) Stress-migration related electromigration damage mechanism in passivated, narrow interconnects. Appl Phys Lett 59:1464
45. Ogawa ET, Mcpherson JW, Rosal JA (2002) Stress-induced voiding under vias connected to wide Cu metal leads. In: Proceedings of 40th Annual IEEE international reliability physics symposium (IRPS), pp 312–321
46. Wang TC, Hsieh TE, Wang M-T, Su D-S, Chang C-H, Wang YL, Lee JY-M (2005) Stress migration and electromigration improvement for copper dual damascene interconnection. J Electrochem Soc 152:45–49
47. Hommel M, Fischer AH, Glasow AV, Zitzelsberger AE (2002) Stress-induced voiding in aluminum and copper interconnects. In: Stress-induced phenomena in metallization: Sixth international workshop on stress-induced phenomena in metallization, pp 157–168
48. An JH, Ferreira PJ (2006) In situ transmission electron microscopy observations of 1.8 μm and 180 nm Cu interconnects under thermal stresses. Appl Phys Lett 89:151919
49. Fayolle M, Passemard G, Assous M, Louis D, Beverina A, Gobil Y, Cluzel J, Arnaud L (2002) Integration of copper with an organic low-k dielectric in 0.12 μm node interconnect. Microelectronic Eng 60(1–2):119–124
50. Laconte J, Iker F, Jorez S, Andre N, Proost J, Pardoen T, Flandre D, Raskin JP (2004) Thin films stress extraction using micromachined structures and wafer curvature measurements. Microelectron Eng 76(1–2):219–226
51. Rivero C, Gergaud P, Gailhanou M, Thomas O, Froment B, Jaouen H, Carron V (2005) Combined synchrotron x-ray diffraction and wafer curvature measurements during Ni-Si reactive film formation. Appl Phys Lett 87:1–3
52. Budiman AS, Nix WD, Tamura N, Valek BC, Gadre K, Maiz J, Spolenak R, Patel JR (2006) Crystal plasticity in Cu damascene interconnect lines undergoing electromigration as revealed by synchrotron x-ray microdiffraction. Appl Phys Lett 88:233515
53. Rhee SH, Du Y, Ho PS (2003) Thermal stress characteristics of Cu/oxide and Cu/low-k submicron interconnect structures. J Appl Phys 93:3926–3933
54. Igic PM, Mawby PA (1999) An advanced finite element strategy for thermal stress field investigation in aluminium interconnections during processing of very large scale integration multilevel structures. Microelectron J 30:1207–1212
55. Shen Y-L (2006) Thermo-mechanical stresses in copper interconnects: a modeling analysis. Microelectron Eng 83:446–459
56. Paik J-M, Park H, Joo Y-C (2004) Effect of low-k dielectric on stress and stress-induced damage in Cu interconnects. Microelectron Eng 71:348–357
57. Flinn PA (1995) Mechanical stress in VLSI interconnections: origins, effects, measurement, and modeling. Mater Res Bull 20:70–73
58. Shi LT, Tu KN (1994) Finite-element modeling of stress distribution and migration in interconnecting studs of a three-dimensional multilevel device structure. Appl Phys Lett 65:1516

59. Ho PS, Yeo IS, Liao CN, Anderson SGH, Kawasaki H (1995) Thermal stress and relaxation behaviour of Al(Cu) submicroninterconnects. In: 4th International conference on solid-state and integrated circuit technology, Beijing, China, pp 408–412

60. Niwa H, Yagi H, Tsuchikawa H, Masaharu K (1990) Stress distribution in an aluminum interconnect of very large scale integration. J Appl Phys 68:328–333

61. Valek BC, Bravman JC, Tamura N, MacDowell AA, Celestre RS, Padmore HA, Spolenak R, Brown WL, Batterman BW, Patel JR (2002) Electromigration-induced plastic deformation in passivated metal lines. Appl Phys Lett 81:4168

62. Hou Y, Tan CM (2007) Blech effect in Cu interconenects with oxide and low-k dielectrics. In: 14th International symposium on the physics and failure analysis of integrated circuits, IEEE, Bangalore, India, p 65

63. Paik J-M, Park H, Joo Y-C, Park K-C (2005) Effect of dielectric materials on stress-induced damage modes in damascene Cu lines. J Appl Phys 97:104513

64. Korhonen MA, Black RD, Li C-Y (1991) Stress relaxation of passivated aluminum line metallizations on silicon substrates. J Appl Phys 69:1748–1755

65. Wikstrom A, Gudmundson P, Suresh S (1999) Analysis of average thermal stresses in passivated metal interconnects. J Appl Phys 86:6088–6095

66. Chang CW, Thompson CV, Gan CL, Pey KL, Choi WK, Lim YK (2007) Effects of microvoids on the linewidth dependence of electromigration failure of dual-damascene copper interconnects. Appl Phys Lett 90:193505

67. Shen Y-L, Guo YL, Minor CA (2000) Voiding induced stress redistribution and its reliability implications in metal interconnects. Acta Mater 48:1667–1678

68. Thouless MD, Rodbell KP, Cabral C (1996) Effect of a surface layer on the stress relaxation of thin films. J Vacuum Sci Technol A 14:2454

69. Keller R-M, Baker SP, Arzt E (1998) Quantitative analysis of strengthening mechanisms in thin Cu films: effects of film thickness, grain size and passivation. J Mater Res 13:1307–1317

70. Suo Z (2003) Reliability of interconnect structures. In: Gerberich W, Yang W (eds) Comprehensive structural integrity, vol 8. Elsevier, Amsterdam, pp 265–324

71. Hou Y, Tan CM (2008) Stress-induced voiding study in integrated circuit interconnects. Semicond Sci Technol 23:075023–075031

72. Korhonen MA, Black RD, Li C-Y (1993) Stress evolution due to electromigration in confined metal lines. J Appl Phys 73:3790–3799

73. Zhai CJ, Blish RC (2005) A physically based lifetime model for stress-induced voiding in interconnects. J Appl Phys 97:113503

74. Fischer AH, Zitzelsberger AE (2001) The quantitative assessment of stress-induced voiding in process qualification. In: Proceedings of 39th IEEE/IRPS conference, Orlando, Florida, IEEE, New York, pp 334–340

75. Tan CM, Hou Y (2007) Lifetime modeling for stress-induced voiding in integrated circuit interconnections. Appl Phys Lett 91:061904

76. Ogawa ET, McPherson JW, Rosal JA, Dickerson KJ, Chiu T-C, Tsung LY, Jain MK, Bonifield TD, Ondrusek JC, Mckee WR (2002) Stress-induced voiding under vias connected to wide Cu metal leads. In: Proceedings of 40th IEEE/IRPS conference, Dallas, Texas, IEEE, New York, pp 312–331

77. Aoyagi M, Asada K (1999) Vacancy distribution in aluminum interconnections on semiconductor devices. Jpn J Appl Phys 38:1909–1914 Part 1 (Regular Papers, Short Notes & Review Papers)

78. Aoyagi M (2003) Modeling of vacancy flux due to stress-induced migration. J Vacuum Sci Technol B Microelectron Nanometer Struct 21:1314–1317

79. Aoyagi M (2006) Change in electrical resistance caused by stress-induced migration. J Vacuum Sci Technol B Microelectron Nanometer Struct 24:250–254

80. Reimbold G, Sicardy O, Arnaud L, Fillot F, Torres J (2002) Mechanical stress measurements in damascene copper interconnects and influence on electromigration parameters. IEDM Tech Digest 745–748

81. Bruynseraede C, Tokei Z, Iacopi F, Beyer GP, Michelon J, Maex K (2005) The impact of scaling on electromigration reliability. In: Proceedings of 43rd IEEE/IRPS conference, pp 7–17
82. Suo Z (2003) Interfacial and Nanoscale Failure. In: Gerberich W, Yang W (eds) Reliability of interconnect structures. Comprehensive Structural Integrity (Milne I, Ritchie RO, Karihaloo B, Editors-in-Chief), 8:265–324
83. Gan ZH, Shao W, Mhaisalkar SG, Chen Z, Gusak A (2006) Experimental and numerical studies of stress migration in Cu interconnects embedded in different dielectrics. In: Stress-induced phenomena in metallization. Eighth international workshop on stress-induced phenomena in metallization, AIP, vol 817, pp 269–274
84. Hau-Riege CS, Hau-Riege SP, Marathe AP (2004) The effect of interlevel dielectric on the critical tensile stress to void nucleation for the reliability of Cu interconnects. J Appl Phys 96:5792–5796
85. Tan CM, Hou Y, Li W (2007) Revisit to the finite element modeling of electromigration for narrow interconnects. J Appl Phys 102:033705
86. Sharma P, Ganti S, Ardebili H, Alizadeh A (2004) On the scaling of thermal stresses in passivated nanointerconnects. J Appl Phys 95:2763–2769
87. Chiras S, Charke DR (2000) Dielectric cracking produced by electromigration in microelectronic interconnects. J Appl Phys 88:6302–6312
88. Atrash F, Sherman D (2006) Analysis of the residual stresses, the biaxial modulus, and the interfacial fracture energy of low-k dielectric thin films. J Appl Phys 100:103510–103517
89. Du Y, Wang G, Merrill C, Ho PS (2002) Thermal stress and debonding in Cu/low-k damascene line structures. In: 52nd Electronic components and technology conference, pp 859–864
90. Shen Y-L, Ramamurty U (2003) Temperature-dependent inelastic response of passivated copper films: experiments, analyses, and implications. J Vacuum Sci Technol B 21:1258–1264

Chapter 3
Introduction and General Theory of Finite Element Method

3.1 History of Finite Element Method (FEM)

The complexity of the physics of electromigration and the stress-induced voiding can be seen in Chap. 2. In order to model the physics realistically in today's interconnects so as to obtain better understanding of the physics and to identify key parameters, Finite Element Method (FEM) is needed as pointed out in Chap. 2. To begin the discussion on FEM applications in electromigration and stress-induced voiding, let us have a basic understanding on the FEM. For readers familiar with FEM, this chapter may be skipped.

The inherent physics of most engineering problems can generally be modeled mathematically using partial differential equations (PDE) and integral equations. Analytical solutions to such mathematical models can be obtained by making some simplifications and assumptions in some cases. However, for a more realistic modeling, either the geometry or certain features of the problem will be irregular or "arbitrary," and no simple analytical solution can be found for these models, and numerical solutions instead of the exact closed-form solutions will be sought. Finite element method (FEM) (or sometimes referred as finite element analysis —FEA) was developed as a powerful numerical analysis technique for obtaining approximate solutions applicable to such problems and hence it can be applied to a wide range of engineering problems.

Finite element method was originally developed for structural/stress analysis in complex civil and aeronautical structures. By the early 1970s, FEM was limited to expensive mainframe computers generally owned by the aeronautics, automotive, defense, and nuclear industries. As computers are becoming affordable and their computation power is increasing tremendously, the application of FEM is extended to academic research and almost all engineering disciplines successfully, with civil, mechanical, and aerospace engineers as the most frequent users of the method. The areas of applications of FEM include heat transfer, fluid mechanics,

C. M. Tan et al., *Applications of Finite Element Methods for Reliability Studies on ULSI Interconnections*, Springer Series in Reliability Engineering, DOI: 10.1007/978-0-85729-310-7_3, © Springer-Verlag London Limited 2011

electromagnetism, biomechanics, geomechanics, acoustics, etc. FEM is also good for problem solving with arbitrary (complicated or irregular) geometries. A more exciting merit of FEM is its application to multidisciplinary problems where there is a coupling between two or more of the disciplines, such as the thermal stress analysis in microelectronic structures where there is a natural coupling between heat transfer and displacements (and thus stresses), as well as aeroelasticity where there is a strong coupling between external flow and the distortion of the wing. Because of its diversity and flexibility as an analysis tool, FEM is receiving increasing attention in academic as well as in industry today. Table 3.1 gives a summary of the history of FEM.

3.2 Principle of FEM

The field of mechanics can be classified into three major areas, namely Theoretical Mechanics, Applied Mechanics, and Computational Mechanics. Theoretical mechanics deals with fundamental laws and principles of mechanics. Applied mechanics are the applications of the theoretical knowledge to scientific and engineering problems, where a critical procedure is to construct mathematical models for the physical phenomena. Computational mechanics solves specific scientific and engineering problems through numerical methods implemented in computers. The FEM was originated from the need for such numerical analysis in computational mechanics. There are numerous textbooks introducing the FEM [15, 16].

The basic idea of the FEM is to discretize the domain under investigation into a finite number of elements. The entire domain is thus approximated by the assemblage of these discrete elements. In each individual element, equations with unknown field variables are then developed on the basis of the governing equations (as PDE or DE) from mathematical model by assuming a shape function (sometimes known as interpolation function) to approximate the physical behavior of an element. The shape functions are defined in terms of the values of the field variables at specified points called nodes. Nodes are usually located on the element boundaries where adjacent elements are connected. In addition to boundary nodes, an element may also have a few interior nodes. The nodal values of the field variable and the interpolation functions for the elements completely define the behavior of the field variable within the elements. This procedure will be discussed in detail later in next section. Figure 3.1 gives a schematic for discretization of the domain of interest into elements/nodes as a 2D example.

With appropriate boundary conditions, initial conditions, and loading applied to the nodes/elements, the equations of all the elements are combined together and solved simultaneously to obtain the basic nodal results, such as displacement for structural analysis or temperature for heat transfer problem. These nodal results are called degree of freedom (DOF). Based on the DOF values at each node, other derivatives of the DOF are possible, such as principal stresses in structural analysis

Table 3.1 History of FEM

Time period	Description
1950–1960	Initiated for structural analysis in aeronautical and civil engineering. First developed in 1943 by Courant [1], who utilized the Ritz method of numerical analysis and minimization of variational calculus to obtain approximate solutions to vibration systems. Shortly thereafter, a paper published in 1956 by Turner et al. established a broader definition of numerical analysis [2]. All these pioneers were in the aerospace industry since only they had the capability to get mainframe computers at that time
1960–1970	Variational formulation approach (in contrast to the previous direct formulation approach) was put forward. This approach was adopted to describe the physical phenomena in terms of minimization of total energy associated with the problem of interest. Earlier workers used finite elements as idealizations of structural components. However, in this decade, a two-step interpretation emerges where discrete elements are used to approximate continuum so as to approximate real structures
	NASA proposed the development of the finite element software NASTRAN in 1965
1970–1980	Textbooks on finite element analysis became available. Commercial FEM codes gradually gained importance, and they provided the capability to check how the real world works through numerical solutions with complex and high order elements [3]
	ANSYS, Inc. was established in 1970. ANSYS is an engineering simulation software provider, that develops general purpose finite element analysis and computational fluid dynamics software. In 1971, ANSYS FEM software was released for the first time
	ABAQUS, Inc. was established in 1978. ABAQUS is a commercial software package for finite element analysis developed by SIMULIA
1980–1990	The application of FEM was extended to other engineering fields, such as the electromagnetic and fluid dynamics problems. As an example, a book entitled "Finite Elements for Electrical Engineers" was published by Silvester and Ferrari in 1983
1990–2010	More books on FEM application in electrical engineering fields were published. To list a few:
	J. Jin, The FEM in Electromagnetics, 1st edn. 1993, Wiley
	S. J. Salon, Finite Element Analysis of Electrical Machines, 1995, Springer
	N. Bianchi, Electrical Machine Analysis Using Finite Elements, 2005, CRC
	Since 1990, FEM is being widely applied to understand the underlying physical mechanisms in VLSI reliability such as electromigration [4–8], stress migration [9–12], and dielectric breakdown [13, 14]

Fig. 3.1 A schematic drawing of the discretization of the domain of interest into elements/nodes

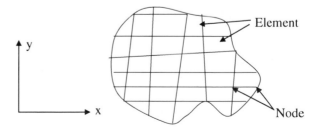

and heat fluxes in heat transfer analysis. Table 3.2 lists the DOF of some analysis involved in this book and its derivative parameters in the main analysis.

3.3 General Procedure of Finite Element Method

To perform FEA without using commercially available software such as ANSYS, ABAQUS or COMSOL, a general step-by-step procedure will be described below. This procedure is hidden in the commercial software and embedded in the Element Types that user selects. However, an understanding of the principles of the finite element calculation as described by the procedure is necessary in order to better interpret the results provided by the software. The procedure is as follows:

(a) The first step is to analyze the system under study, and to derive mathematical equations to describe the system. These equations are called governing equations, and they are usually in the form of differential equation (DE) or PDE.
(b) The second step is to decide the FEM formulation methodology. There are basically three different approaches to formulate the properties of an individual element.

3.3.1 Direct Approach

In this approach, physical concept (e.g., force equilibrium, energy conservation, etc.) is applied directly to the discretized elements. It is easy in its physical interpretation. The stiffness matrix can be manually traced from a physics point of view. An example is given in Fig. 3.2 for a simple two-force truss member.

Table 3.2 DOF of some analysis involved in this book and its derivative parameters

	DOF	Derivative parameters
Structural analysis	Displacement in x, y, and z directions	Strains; normal stresses; principal stresses; von Mises stress; hydrostatic stress
Heat transfer/ conduction analysis	Temperature	Temperature gradient; thermal flux; heat flow rate
Electric field analysis	Voltage (or current)	Electric field; electric flux density; current density, current density gradient

Fig. 3.2 A simple example of the two-force truss members

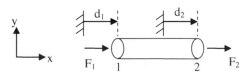

From the definition of basic mechanics, the 1D strain (ε) and stress (σ) of the system are expressed as:

$$\varepsilon = \frac{d_2 - d_1}{L} \tag{3.1}$$

$$\sigma = \frac{F}{A} \tag{3.2}$$

where d_1 and d_2 are the displacement at Node 1 and Node 2 of the truss element, respectively; L and A are the length and the cross-sectional area of the truss, respectively. F is the applied force, and $F = F_2 = -F_1$.

According to the Hooke's law:

$$\sigma = E\varepsilon \tag{3.3}$$

and combining Eqs. 3.1–3.3, we have

$$k(d_2 - d_1) = F \tag{3.4}$$

where

$$k = \frac{EA}{L}. \tag{3.5}$$

Since $F = F_2 = -F_1$, Eq. 3.4 can be further expressed in terms of F_1 and F_2 as follows:

$$k(d_1 - d_2) = F_1 \tag{3.6}$$

$$k(-d_1 + d_2) = F_2 \tag{3.7}$$

Equations 3.6 and 3.7 can be transformed into matrix equation given as:

$$k \begin{bmatrix} 1 & -1 \\ -1 & 1 \end{bmatrix} \begin{Bmatrix} d_1 \\ d_2 \end{Bmatrix} = \begin{Bmatrix} F_1 \\ F_2 \end{Bmatrix} \tag{3.8}$$

or

$$[K]\{d\} = \{F\} \tag{3.9}$$

where

$$[K] = k \begin{bmatrix} 1 & -1 \\ -1 & 1 \end{bmatrix} \tag{3.10}$$

The matrix $[K]$ is called the stiffness matrix.

Given the element stiffness matrix $[K]$ and element nodal load vector $\{F\}$, the element nodal displacement vector $\{d\}$ can thus be solved using the element matrix equation of (3.9) as follows:

$$\{d\} = [K]^{-1}\{F\} \tag{3.11}$$

From this simple example, the concepts of node (Node 1 and Node 2) and element (i.e., the truss element) are illustrated. The displacement $\{d\}$ is the DOF, which is an independent parameter of response from the load $\{F\}$.

Further examples can be found in the book by Moaveni [17]. This approach is easy to understand when using the FEM for the first time. Basically this approach is applicable to relatively simple problems.

3.3.2 Variational Approach

Besides the PDE/DE mentioned above, physical phenomena can be described in terms of minimization of total energy (or functional) associated with the problem, and such approach is called "variational formulation." Finite element formulation can be derived using this variational formulation as long as there exists a variational principle corresponding to the problem of interest. An example is a very important physical principle to describe a deformation process of an elastic body, namely the Principle of Minimum Total Potential Energy, which can be summarized as follows.

The total potential energy of an elastic body is given in

$$E = X + Y : \text{ Total Potential Energy} \qquad (3.12)$$

where X is strain energy, and Y is potential energy, they are from external loads. E is the minimum with respect to the state variables or function variables at equilibrium state.

As the total potential energy is a function of state variables or a function of functions, which is called "functional," the next task is to find the minimum through PDE given as

$$\frac{\partial E}{\partial u_i} = 0, \quad i = 1, \ldots, n. \qquad (3.13)$$

By solving these series of equations, the values of the state variables at minimum E will be available.

While the direct approach can be used to formulate element properties for only the simplest element shapes, the variational approach can be employed for both simple and complex element shapes. The minimum total potential energy formulation is a common approach for generating finite element models in solid mechanics.

3.3.3 Weighted Residual Approach

In this approach, a weighting function (ψ) is assumed. The most popular weighting function is obtained using the Galerkin method developed by Russian mathematician

Boris Galerkin (webpage: http://en.wikipedia.org/wiki/Boris_Galerkin). In this method, ψ is defined in the form of Lagrange interpolation polynomials $L(\varphi) = 0$ with φ as unknown function to be solved.

An example of a 1D boundary-value problem is given below for easier understanding of the weighted residual approach. Considering a general form of a governing DE:

$$\frac{d}{dx}\left[-\alpha(x) \cdot \frac{d\varphi}{dx}\right] + \beta(x) \cdot \varphi - f(x) = 0, \quad x_1 \leq x \leq x_2 \tag{3.14}$$

with the following boundary conditions:

$$\varphi(x_1) = \varphi_1 \quad \text{or} \quad \left[-\alpha(x) \cdot \frac{d\varphi}{dx}\right]_{x1} = u_1 \quad \text{at } x = x_1$$

$$\varphi(x_2) = \varphi_2 \quad \text{or} \quad \left[-\alpha(x) \cdot \frac{d\varphi}{dx}\right]_{x2} = u_2 \quad \text{at } x = x_2$$

where x is the independent variable; $x_1 \leq x \leq x_2$ is the domain; φ is the unknown function to be solved; α, β are the material or physical properties of the system; $f(x)$ is the interior load; $u = -\alpha(x) \cdot d\varphi/dx$ is the flux; $d\varphi/dx$ is gradient of the unknown; φ_1, φ_2, u_1, u_2 is the boundary loads (with known values).

The corresponding trial solution to approximate the exact solution of Eq. 3.14 is:

$$\tilde{\varphi}(x, a) = \sum_{j=1}^{n} a_j \phi_j(x) \tag{3.15}$$

where $\phi_j(x)$ is the trial functions, which are known as polynomials. Therefore, the purpose is to solve and get the generalized coordinates/coefficients a_j, which can be derived in the following way.

$$\frac{d\tilde{\varphi}(x, a)}{dx} = \sum_{j=1}^{n} a_j \frac{d\phi_j(x)}{dx} \tag{3.16}$$

Since $\tilde{\varphi}(x, a)$ in Eq. 3.15 is used to approximate the φ in Eq. 3.14, there would be a residual from Eq. 3.14 given by

$$R(x, a) = -\frac{d}{dx}\left[\alpha(x) \cdot \frac{d\tilde{\varphi}(x, a)}{dx}\right] + \beta(x) \cdot \tilde{\varphi}(x, a) - f(x). \tag{3.17}$$

Now the principle behind this weighted residual approach is to minimize the residual over a domain (Ω) by setting it to zero, as defined in the following equation:

$$\int_{\Omega} \psi L(\varphi) d\Omega = 0. \tag{3.18}$$

Substitute the residual $R(x, a)$ as expressed in Eq. 3.17 to replace the $L(\varphi)$ in Eq. 3.18, and the weighting function (ψ) is replaced by the trial function ϕ_i associated with each of the coefficient $\cdot a_i$, Eq. 3.18 becomes

$$\int_{x1}^{x2} R(x, a)\phi_i(x)dx$$

$$= \int_{x1}^{x2} \left(-\frac{d}{dx}\left[\alpha(x) \cdot \frac{d\tilde{\varphi}(x, a)}{dx}\right] + \beta(x) \cdot \tilde{\varphi}(x, a) - f(x) \right) \phi_i(x)dx = 0, \quad i = 1, 2, \ldots, n$$

(3.19)

The coefficient a_j in Eq. 3.15 can be obtained by minimizing the residual in the average sense over the domain as expressed in Eq. 3.19. Using integration by parts, we have

$$\int_{x1}^{x2} \left[\alpha \cdot \frac{d\tilde{\varphi}}{dx} \cdot \frac{d\phi_i}{dx} + \beta \cdot \tilde{\varphi} \cdot \phi_i \right] \cdot dx$$

$$= \int_{x1}^{x2} f \cdot \phi_i \cdot dx - \left[\left(-\alpha \cdot \frac{d\tilde{\varphi}}{dx} \right) \cdot \phi_i \right]_{x1}^{x2}.$$

(3.20)

$$= \int_{x1}^{x2} f \cdot \phi_i \cdot dx - [u \cdot \phi_i]_{x1}^{x2}, \quad i = 1, 2, \ldots, n$$

By substituting Eq. 3.16 into Eq. 3.20, we have

$$\int_{x1}^{x2} \left[\alpha \cdot \left(\sum_{j=1}^{n} a_j \frac{d\phi_j}{dx} \right) \cdot \frac{d\phi_i}{dx} + \beta \cdot \left(\sum_{j=1}^{n} a_j\phi_j \right) \cdot \phi_i \right] \cdot dx$$

$$= \int_{x1}^{x2} f \cdot \phi_i \cdot dx - [u \cdot \phi_i]_{x1}^{x2}, \quad i = 1, 2, \ldots, n$$

$$\Rightarrow \sum_{j=1}^{n} \left[\int_{x1}^{x2} \left[\alpha \cdot \frac{d\phi_j}{dx} \cdot \frac{d\phi_i}{dx} + \beta \cdot \phi_j \cdot \phi_i \right] \cdot dx \right] a_j$$

$$= \int_{x1}^{x2} f \cdot \phi_i \cdot dx - [u \cdot \phi_i]_{x1}^{x2}, \quad i = 1, 2, \ldots, n$$

$$\Rightarrow \begin{bmatrix} K_{11} & K_{12} & \cdots & K_{1n} \\ K_{21} & K_{22} & \cdots & K_{2n} \\ \cdots & \cdots & \cdots & \cdots \\ K_{n1} & K_{n2} & \cdots & K_{nn} \end{bmatrix} \begin{Bmatrix} a_1 \\ a_2 \\ \cdots \\ a_n \end{Bmatrix} = \begin{Bmatrix} F_1 \\ F_2 \\ \cdots \\ F_n \end{Bmatrix}$$

(3.21)

where the terms in the stiffness matrix are given by

$$K_{ij} = \int_{x1}^{x2} \left[\alpha \cdot \frac{d\phi_j}{dx} \cdot \frac{d\phi_i}{dx} + \beta \cdot \phi_j \cdot \phi_i \right] \cdot dx \qquad (3.22)$$

and the terms in the load matrix are given by

$$F_i = \int_{x1}^{x2} f \cdot \phi_i \cdot dx - [u \cdot \phi_i]_{x1}^{x2} \qquad (3.23)$$

Equation 3.21 is similar to Eq. 3.8, and with Eq. 3.21, a_i can be solved.

A more specific example is given below. Here, we have the temperature distribution $T(r)$ near the core of a reaction tube approximated by the following 1D governing equation

$$\frac{d}{dr} \left[(r+1) \cdot \frac{dT}{dr} \right] = 0, \quad (1 \le x \le 2) \qquad (3.24)$$

where r is the distance from the center of the tube. The boundary conditions for Eq. 3.24 are

$$T(1) = 1 \quad \text{and} \quad q(2) = 1$$

where $q(r)$ is the heat flux given by

$$q(r) = -(r+1) \cdot \frac{dT}{dr}. \qquad (3.25)$$

The independent variable is r, and the unknown to be solved is $T(r)$. Comparing Eq. 3.24 with Eq. 3.14, we have the following corresponding parameters:

$$\alpha(r) = -(r+1); \quad \beta(r) = 0; \quad f(r) = 0; \quad r_1 = 1; \quad r_2 = 2.$$

Thus, for this example, Eq. 3.19 can be expressed as:

$$\int_1^2 \phi_i \cdot \frac{d}{dr} \left[(r+1) \cdot \frac{dT}{dr} \right] \cdot dr, \quad i = 1, 2, \ldots, n \qquad (3.26)$$

In order to solve Eq. 3.26, we need to choose a trial solution first as follows:

$$T = b_0 + b_1 r + b_2 r^2 \qquad (3.27)$$

and then

$$\frac{dT}{dr} = b_1 + 2b_2 r \qquad (3.28)$$

and

$$q = -(r + 1) \cdot (b_1 + 2b_2 r) \tag{3.29}$$

Applying the boundary conditions to Eq. 3.27, we have

$$T(1) = b_0 + b_1 + b_2 = 1, \quad \text{i.e., } b_0 = 1 - b_1 - b_2 \tag{3.30}$$

$$q(2) = -(2 + 1) \cdot (b_1 + 4b_2) = 1, \quad \text{i.e., } b_1 = -\frac{1}{3} - 4b_2 \tag{3.31}$$

Substitute Eqs. 3.30 and 3.31 into Eq. 3.27, we have

$$\begin{aligned} T &= \left(\frac{4}{3} - \frac{1}{3}r\right) + b_2\left(3 - 4r + r^2\right) \\ &= \phi_1(r) + b_2\phi_2(r) \end{aligned} \tag{3.32}$$

where

$$\phi_1(r) = \frac{4}{3} - \frac{1}{3}r \tag{3.33}$$

$$\phi_2(r) = 3 - 4r + r^2 \tag{3.34}$$

Equation 3.15 will now take the following form for this example:

$$\tilde{T}(r, a) = \sum_{j=1}^{n} a_j \phi_j(r) = a_1\phi_1(r) + a_2\phi_2(r) \tag{3.35}$$

with $a_1 = 1$ and $a_2 = b_2$.

The next task is to get the value for a_2 (or b_2 in Eq. 3.32). After taking differentiation on Eqs. 3.33 and 3.34, we have

$$\frac{d\phi_1}{dr} = -\frac{1}{3} \tag{3.36}$$

$$\frac{d\phi_2}{dr} = -4 + 2r \tag{3.37}$$

Therefore, the differentiation of Eq. 3.32 gives:

$$\frac{dT}{dr} = -\frac{1}{3} + 2b_2(r - 2) \tag{3.38}$$

Using the Galerkin method, Eq. 3.19 is expressed in this example as:

$$\int_{1}^{2} \left[-(r + 1)\frac{dT}{dr}\frac{d\phi_i}{dr}\right] dr - \left[-(r + 1)\frac{dT}{dr} \cdot \phi_i\right]_{1}^{2} = 0 \tag{3.39}$$

Since only b_2 is needed to be solved, taking $i = 2$, we have

$$\int_1^2 \left[-(r+1)\frac{dT}{dr}\frac{d\phi_2}{dr} \right] dr - \left[-(r+1)\frac{dT}{dr} \cdot \phi_2 \right]_1^2 = 0$$

$$\Rightarrow \int_1^2 \left[-(r+1)\left(-\frac{1}{3} + 2b_2(r-2) \right)(-4+2r) \right] dr$$

$$- \left[-(r+1)\left(-\frac{1}{3} + 2b_2(r-2) \right)(3 - 4r + r^2) \right]_1^2 = 0$$

$$\Rightarrow b_2 = \frac{2}{27}$$

$$\Rightarrow T(r) = \frac{1}{27}\left(42 - 17r + 2r^2 \right) \tag{3.40}$$

From the calculus, the analytical solution of Eq. 3.24 and the given conditions is

$$T(r) = 1 - \ln\left(\frac{r+1}{2} \right) \tag{3.41}$$

Compare the values of T in terms of r as computed from Eqs. 3.40 and 3.41, respectively, we have the results summarized in Table 3.3. From the table, one can see that both results are comparable to each other.

(c) The third step is to discretize the domain of interest into elements, which also include their corresponding nodes.
(d) The fourth step is to introduce an approximation of the field variable over an element. The idea behind the FEM is that the field variable over an element is represented by its nodal values, and defined by the interpolation functions. An example of the interpolation function is shown in Fig. 3.3 with $n = 4$.

Table 3.3 The comparison results between Eqs. 3.40 and 3.41

r	T_analytical	T_Galerkin
1	1.0000	1.0000
1.1	0.9512	0.9526
1.2	0.9047	0.9067
1.3	0.8602	0.8622
1.4	0.8177	0.8193
1.5	0.7769	0.7778
1.6	0.7376	0.7378
1.7	0.6999	0.6993
1.8	0.6635	0.6622
1.9	0.6284	0.6267
2	0.5945	0.5926

Fig. 3.3 Definition of the
element interpolation
function

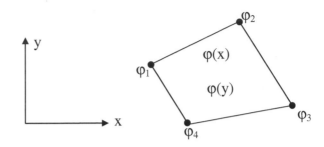

$$\varphi(x) = \sum_{j=1}^{n} \varphi_j(x) N_j(x)$$

$$= N_1(x)\varphi_1(x) + N_2(x)\varphi_2(x) + N_3(x)\varphi_3(x) + N_4(x)\varphi_4(x) \qquad (3.42)$$

$$\varphi(y) = \sum_{j=1}^{n} \varphi_j(y) N_j(y)$$

$$= N_1(y)\varphi_1(y) + N_2(y)\varphi_2(y) + N_3(y)\varphi_3(y) + N_4(y)\varphi_4(y) \qquad (3.43)$$

Here $\varphi(x)$ and $\varphi(y)$ are the element values in x and y direction, respectively; φ's are the nodal values of the field variables, and N_j is the interpolation function. In other words, the a_j and $\phi_j(x)$ in Eq. 3.15 are replaced by φ_j and N_j, respectively.

Examples of φ_j are the nodal values of the field variable like the DOF as shown in Table 3.2. Table 3.4 gives examples of important shape functions used in ANSYS. More complete shape functions are given in the manuals of finite element software [18].

(e) The fifth step is to evaluate Eqs. 3.14–3.23 in each element by substituting the above interpolation function into the integrals (Eq. 3.18) in Step 2, and obtain the following element polynomial equations for numerical evaluation:

$$[K]^e \{\varphi\}^e = \{f\}^e \qquad (3.44)$$

where $[K]$ is the stiffness matrix, $\{f\}$ is the load matrix, and the superscript "e" represents element. This corresponds to the simplified case shown in Eq. 3.21 where the a_j's have been replaced by φ_j's.

Figure 3.4 is an example of three elements each consisting of two nodes. Equation 3.44 for Elements (1–3) are given as

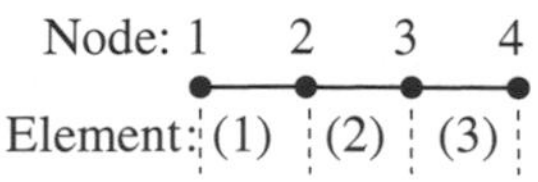

Fig. 3.4 Three elements
each consisting of two nodes

Node: 1 2 3 4

Element: (1) (2) (3)

Table 3.4 Examples of shape functions used in ANSYS

Element type	Geometrical representation	Node	DOF	Shape function
2D line (e.g., LINK1)		I, J	u, v (e.g., displacement in x and y)	$u = u_I \cdot \frac{1-s}{2} + u_J \cdot \frac{1+s}{2}$ $v = v_I \cdot \frac{1-s}{2} + v_J \cdot \frac{1+s}{2}$
2D 4-node quadrilateral solid (e.g., PLANE42)		I, J, K, L	u, v (e.g., displacement in x and y)	$u = u_I \cdot \frac{(1-s)(1-t)}{4} + u_J \cdot \frac{(1+s)(1-t)}{4}$ $\quad + u_K \cdot \frac{(1+s)(1+t)}{4} + u_L \cdot \frac{(1-s)(1+t)}{4}$ $v = v_I \cdot \frac{(1-s)(1-t)}{4} + v_J \cdot \frac{(1+s)(1-t)}{4}$ $\quad + v_K \cdot \frac{(1+s)(1+t)}{4} + v_L \cdot \frac{(1-s)(1+t)}{4}$

$$\begin{bmatrix} K_{11} & K_{12} \\ K_{21} & K_{22} \end{bmatrix}^{(1)} \left\{ \begin{matrix} \varphi_1 \\ \varphi_2 \end{matrix} \right\} = \left\{ \begin{matrix} F_1 \\ F_2 \end{matrix} \right\}^{(1)} \tag{3.45}$$

$$\begin{bmatrix} K_{22} & K_{23} \\ K_{32} & K_{33} \end{bmatrix}^{(2)} \left\{ \begin{matrix} \varphi_2 \\ \varphi_3 \end{matrix} \right\} = \left\{ \begin{matrix} F_2 \\ F_3 \end{matrix} \right\}^{(2)} \tag{3.46}$$

$$\begin{bmatrix} K_{33} & K_{34} \\ K_{43} & K_{44} \end{bmatrix}^{(3)} \left\{ \begin{matrix} \varphi_3 \\ \varphi_4 \end{matrix} \right\} = \left\{ \begin{matrix} F_3 \\ F_4 \end{matrix} \right\}^{(3)} \tag{3.47}$$

(f) The sixth step is to combine all the element equations to get the global matrix as follows.

$$[K]\{\varphi\} = \{F\} \tag{3.48}$$

The above example (Fig. 3.4) is further elaborated as follows. Assembling Elements (1–3) in Eqs. 3.45–3.47 we get

$$\begin{bmatrix} K_{11}^{(1)} & K_{12}^{(1)} & & \\ K_{21}^{(1)} & K_{22}^{(1)} + K_{22}^{(2)} & K_{23}^{(2)} & \\ & K_{32}^{(2)} & K_{33}^{(2)} + K_{33}^{(3)} & K_{34}^{(3)} \\ & & K_{43}^{(3)} & K_{44}^{(3)} \end{bmatrix} \left\{ \begin{matrix} \varphi_1 \\ \varphi_2 \\ \varphi_3 \\ \varphi_4 \end{matrix} \right\} = \left\{ \begin{matrix} F_1^{(1)} \\ F_2^{(1)} + F_2^{(2)} \\ F_3^{(2)} + F_3^{(3)} \\ F_4^{(3)} \end{matrix} \right\} \tag{3.49}$$

(g) The seventh step is to solve the global matrix equations to obtain the unknowns (independent variables, DOF) as follows:

$$\{\varphi\} = [K]^{-1}\{F\}. \tag{3.50}$$

(h) The derivative parameters as described in Table 3.2 can then be computed as necessary.

3.4 Application Categories of Finite Element Method

There are basically three categories of applications for the FEM, depending on the nature of the problem to be solved. The first category consists of the problems known as equilibrium problems or time-independent problems. Majority of the FEM applications fall into this category. For the solutions of equilibrium problems in solid mechanics area, the displacement and the stress distributions for a given mechanical or thermal loading will be determined. Similarly, for the solutions of equilibrium problems in fluid mechanics, the pressure, velocity, temperature, and density distributions under steady-state conditions will be computed.

The second category includes the so-called eigenvalue problems of solid and fluid mechanics. These are steady-state problems whose solution often requires the determination of natural frequencies and modes of vibration of solids and fluids. Examples of eigenvalue problems involving both solid and fluid mechanics appear in civil engineering when the interaction of lakes and dams is considered, and in aerospace engineering when the sloshing of liquid fuels in flexible tanks is involved. Another class of eigenvalue problems include the stability of structures and the stability of laminar flows.

The third category is the multitude of time-dependent or propagation problems of continuum mechanics. This category consists of the problems of the above-mentioned first two categories with the time dimension added.

As FEM analysis solves numerical models of designs with their constituent materials being stressed, it is widely used in new product design and existing product design refinement. One is able to verify a proposed design for meeting the customers' specifications prior to manufacturing or construction through FEM. In the case of structural failure, FEM can be helpful for root cause isolation and determination of the design modifications.

Generally, two types of analysis are used in industry today, namely 2D and 3D modeling. While 2D modeling is simplified and allows the analysis to be run on a standard computer, it tends to yield less accurate results. The 3D modeling, on the other hand, produces more accurate results while sacrificing the ability to run on all but the fastest computers, effectively. However, with the increasing power of computer today, 3D modeling is increasingly common. For either type of modeling, programmer can insert numerous algorithms (functions) which make the system to behave either linearly or nonlinearly. Linear systems are far less complex and generally do not take into account of the plastic deformation. Nonlinear systems account for plastic deformation, and many are also capable of modeling a material all the way to fracture.

3.5 Commercial Software for Finite Element Method

As described in Sect. 3.3, the procedure to perform a FEA is not straight forward. The key is the finite element formulation, which requires users performing the analysis to have a good understanding of both the theory and mathematics of the FEM. However, with the rapid development of computer hardware and software, it is now possible for one to perform FEM analysis without much mastering the details of the finite element formulation technique. The development of the hardware makes the calculation faster even for an increasingly complicated structure. The availability of commercial FEM software integrates the finite element formulation into each element type, which provides users a "black box" where the

Table 3.5 The basic steps in using a commercial software to perform the finite element analysis

Pre-processing	Build the model
	Create the geometry to represent the domain under study
	Select a suitable Element Type to be used as a "black box" where the underlying physics, mathematical equations, and finite element formulation are hidden
	Discretize the domain into finite elements (i.e., sub-divide the problem into nodes and elements)
	Assign materials properties (attributes) such as thermal conductivity for a thermal analysis and Young's modulus for a stress analysis
	Apply boundary conditions, initial conditions, and loadings
Solution	Solve a set of linear or nonlinear algebraic equations simultaneously (note that the user will not see these equations) to obtain nodal results, i.e., the DOF items listed in Table 3.2
Post-processing	To obtain other important information, for example, one may get the derivative parameters listed in Table 3.2. In addition, one may also obtain the contour plots of the DOF items and derivative parameters

underlying physics, mathematical equations, and finite element formulation are hidden. Several commercial softwares are available today, such as ABAQUS (http://www.simulia.com/), MSC/NASTRAN (http://www.mscsoftware.com/), MSC/MARC (http://www.mscsoftware.com/), ANSYS (http://www.ansys.com/), and RASNA (http://www.ptc.com). Among them, ANSYS and ABAQUS are the most popular ones. Table 3.5 lists the basic steps in using a commercial software to perform FEA.

ANSYS (webpage: http://www.ansys.com/) is a comprehensive general-purpose FEA and computational fluid dynamics software. It is well known for its Mechanical and Multiphysics products. Its computer-aided engineering products have also made it a more powerful and impressive engineering tool for design in industry. ANSYS has a user-friendly multiple windows interface incorporating the Graphical User Interface (GUI), pull-down menus, dialog boxes, and tool bars. ANSYS found its applications in many engineering fields, including aerospace, automotive, and electronics.

ANSYS Mechanical and ANSYS Multiphysics software are FEA tools incorporating pre-processing (geometry creation and meshing), solver, and post-processing modules in a GUI. These are general-purpose finite element modeling packages for solving mechanical problems numerically, including static/dynamic structural analysis (both linear and nonlinear), heat transfer and fluid problems, as well as acoustic and electro-magnetic problems (webpage: http://en.wikipedia.org/wiki/Ansys). ANSYS Multiphysics is also used to solve coupled multidisciplinary problems including thermal-electrical applications (e.g., Joule heating, Seeback effect, Peltier effect, and Thomson effect), structural–thermal applications (e.g., thermo-mechanical stress), and more comprehensive structural–thermal–electrical applications.

ABAQUS (webpage: www.hks.com/) is a commercial high-performance software package for FEA developed by Hibbitt, Karlsson and Sorensen. ABAQUS,

Inc. was established in 1978, and is recently fully transitioned to SIMULIA (webpage: http://www.simulia.com/), the Dassault Systèmes brand for realistic simulation. It enables one to do linear or nonlinear, and static or dynamic types of analysis for a large spectrum of engineering problems. The ABAQUS suite of engineering analysis software packages are used to simulate the physical response of structures and solid bodies to load, temperature, contact, impact, and other environmental conditions. Some of the specific capabilities of ABAQUS are the following: Stress analysis (both static and dynamic responses); Dynamic studies (linear and nonlinear problems); Heat transfer problems; Coupled heat transfer and stress analysis; Coupled pore fluid diffusion and stress analysis problems; Eigen-value buckling prediction; Natural frequency extraction; J-integral evaluation; Geostatic stress state; Dynamic analysis of linear systems by modal methods; Element removal and replacement.

ABAQUS was initially designed to address nonlinear physical behavior, and hence the package has an extensive range of material models. Its elastomeric (rubberlike) material capabilities are particularly noteworthy (webpage: http://en.wikipedia.org/wiki/Abaqus).

3.6 Methodology of Finite Element Method for Interconnect Study

As the scope of this book is on ULSI interconnects, let us now zoom our discussion on the application of FEM on ULSI interconnects. ANSYS will be used to illustrate the methodology.

One popular FEM application for interconnect study is the thermo-mechanical stress analysis. The stress in an interconnect system arising from its non-uniform temperature distribution and the differences in the thermal expansion coefficients of silicon, dielectrics (such as silicon dioxide), and metals (such as Cu). Such an analysis is known as coupled thermo-mechanical analysis, where the thermal analysis is followed by a stress analysis to calculate the thermo-mechanical stress distribution in the system under study. The temperature profile in a system or component is obtained from the thermal analysis, and the temperature profile is then inputted into a stress analysis model as thermal load.

Figure 3.5 gives the procedural flow to perform such coupled-field thermo-mechanical analysis. The first step is to build a model by drawing the geometry to represent the domain under study. The physical environments of the two fields, namely the thermal and mechanical fields are then created by assigning the attributes/element types to the model, and the finite element meshing is performed. For a 3D analysis, 4 node tetrahedral element (Fig. 3.6a) or 8 node brick element (Fig. 3.6b) is generally used. The corresponding higher order elements are 10 node tetrahedral element (Fig. 3.6c) and 20-node brick element (Fig. 3.6d), respectively. The higher order elements are better suited for modeling problems with curved boundaries, at the expense of more computational time.

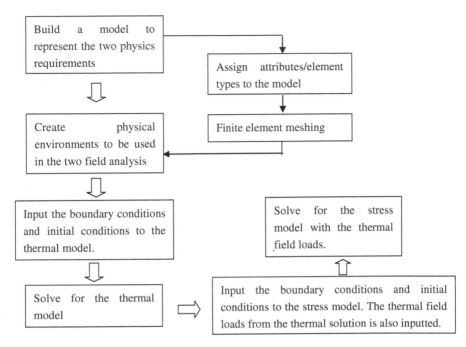

Fig. 3.5 The procedural flow to perform the coupled-field thermo-mechanical analysis

Fig. 3.6 Schematic drawing of elements for 3D analysis: **a** 4 node tetrahedral element; **b** 8 node brick element; **c** 10 node tetrahedral element; and **d** 20 node brick element [18]

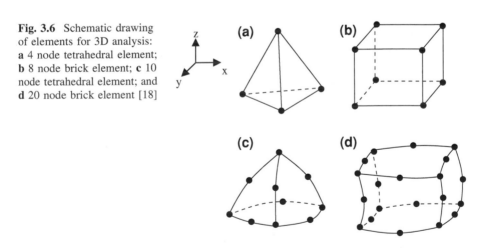

The third step is to read the thermal physics environment as defined by the boundary and initial conditions for the elements. Next is to solve DOF in the domain under study, which will provide the thermal-related nodal solution of the model such as the temperature of each node. After that, the structural physics environment is analyzed, and the boundary and the initial conditions are applied.

In addition, the thermal field loads from the above thermal solution should also be inputted into the structural model. The sixth step is to solve the model in the structural physical environment with the thermal field loads. Detailed examples will be given in the following chapters.

For a better understanding of the coupled-field thermo-mechanical analysis, the following sections provide the details of element selection and theoretical aspects of the thermal and stress analysis.

3.6.1 Thermal Analysis

3.6.1.1 Element Selection for Thermal Analysis

The basis for thermal analysis in ANSYS is a heat balance equation obtained from the principle of conservation of energy. Typical thermal quantities of interest include the temperature distributions, the amount of heat lost or gain, thermal gradients, and thermal fluxes. Elements PLANE55 and SOLID70 are used to analyze the temperature field for 2D and 3D analysis, respectively (Fig. 3.7).

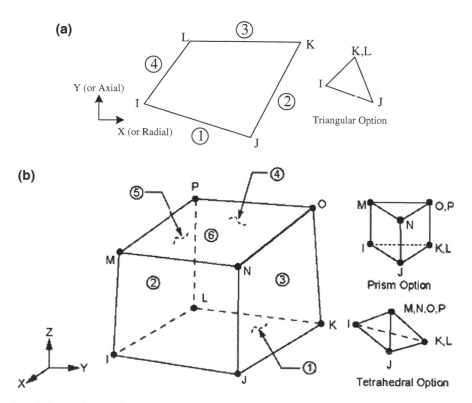

Fig. 3.7 a PLANE55 for 2D and **b** SOLID70 for 3D temperature field analysis [18]

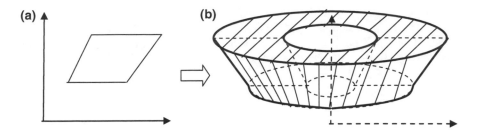

Fig. 3.8 A schematic representation of **a** 2D axisymmetric element for **b** 3D ring

PLANE55 can be used as a plane element or as an axisymmetric ring element with a 2D thermal conduction capability. The 2D axisymmetric element is used to simulate a 3D situation by assuming that each 2D element represents a 3D ring (Fig. 3.8). The element has four nodes with a single DOF (temperature) at each node.

SOLID70 has a 3D thermal conduction capability. The element has eight nodes with a single DOF (temperature) at each node. The element is applicable to a 3D, steady-state or transient thermal analysis.

In order to apply thermal analysis accurately, one needs to understand the theoretical aspect of the thermal analysis, which is detailed in the following section.

3.6.1.2 Theoretical Aspect of Thermal Analysis

Conduction and convection. The first law of thermodynamics states that thermal energy is conserved. The governing equation for the conservation of energy is [18, 19]:

$$\rho c \left(\frac{\partial T}{\partial t} + \{v\}^T \nabla T \right) + \nabla \cdot \{q\} = Q \tag{3.51}$$

where ρ is the density of material; c is the specific heat of material; T is the temperature; t is time;

$$\nabla = \left\lfloor \frac{\partial}{\partial x} \quad \frac{\partial}{\partial y} \quad \frac{\partial}{\partial z} \right\rfloor^T$$

is gradient operator; $\{v\} = \lfloor v_x \quad v_y \quad v_z \rfloor^T$ is the velocity vector for mass transport of heat; $\{q\}$ is the heat flux vector; and Q is heat generation rate per unit volume, including heat source and sink.

Fourier's law of heat conduction [20] is used to relate the heat flux vector to the temperature gradients as follows:

$$\{q\} = -[k]\nabla T \tag{3.52}$$

where

$$[k] = \begin{bmatrix} k_x & 0 & 0 \\ 0 & k_y & 0 \\ 0 & 0 & k_z \end{bmatrix}$$

is thermal conductivity matrix; and k_x, k_y, k_z are thermal conductivity in the element along the x, y, and z directions, respectively.

Combining Eqs. 3.51 and 3.52 yields [18]

$$\rho c \left(\frac{\partial T}{\partial t} + \{v\}^T \nabla T \right) = \nabla \cdot ([k] \nabla T) + Q \tag{3.53}$$

Expanding Eq. 3.53 to its more familiar form in global Cartesian system gives [18]

$$\rho c \left(\frac{\partial T}{\partial t} + v_x \frac{\partial T}{\partial x} + v_y \frac{\partial T}{\partial y} + v_z \frac{\partial T}{\partial z} \right)$$
$$= Q + \frac{\partial}{\partial x} \left(k_x \frac{\partial T}{\partial x} \right) + \frac{\partial}{\partial y} \left(k_y \frac{\partial T}{\partial y} \right) + \frac{\partial}{\partial z} \left(k_z \frac{\partial T}{\partial z} \right) \tag{3.54}$$

Within an interconnection-system, Joule heating is possible when the current passing through the interconnect is sufficient large, which will contribute to the "heat generation" (i.e., $Q \neq 0$). The simulation of Joule heating effects in ANSYS is realized by coupling the electromagnetic and heat transfer analysis. In the case where Joule heating can be ignored when the current passing through it is low, there is no heat generation (i.e., $Q = 0$) and no mass transport for heat (note here that the heat convection on surface is considered as one of the boundary conditions). The temperature governing equation (Eq. 3.54) can then be rewritten and expressed as

$$\rho c \left(\frac{\partial T}{\partial t} \right) = \frac{\partial}{\partial x} \left(k_x \frac{\partial T}{\partial x} \right) + \frac{\partial}{\partial y} \left(k_y \frac{\partial T}{\partial y} \right) + \frac{\partial}{\partial z} \left(k_z \frac{\partial T}{\partial z} \right) \tag{3.55}$$

where the right-hand side of Eq. 3.55 is the heat flux related to thermal conduction.

Two types of boundary conditions are considered for a unit volume in an interconnect as follows:

(1) Specified constant temperature acting over the entire surface:

$$T = T^* \tag{3.56}$$

where T^* is the specified temperature.

(2) Specified convection surfaces acting over the entire surface according to the Newton's law of cooling

$$\{q\}^T \{\eta\} = -h_f \cdot (T_B - T_S) \tag{3.57}$$

Fig. 3.9 Schematic drawing
to illustrate the radiation
coupling between two
surfaces A_i and A_j. where r is
the distance between
differential surfaces i and j;
θ_i (θ_j) is the angle between
N_i (N_j) and the radius line to
surface $d(A_j)$ $(d(A_i))$; N_i (N_j)
is surface normal of $d(A_i)$
$(d(A_j))$ [18]

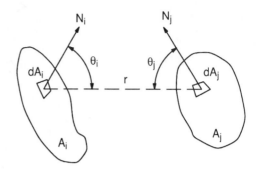

where h_f is the film heat transfer coefficient, T_B is the bulk temperature of adjacent
fluid, and T_S is the temperature at the surface of the model.

Radiation. The above equations consider only conduction and convection.
Another heat transfer mechanism is radiation. Figure 3.9 gives a schematic to
illustrate the radiation coupling between two surfaces i and j. Expanding the
Stefan–Boltzmann law to a two-surface radiation equation, the heat transfer rate
between two surfaces i and j due to radiation is given in [18]:

$$Q_i = \sigma \varepsilon_i F_{ij} A_i \left(T_i^4 - T_j^4 \right) \tag{3.58}$$

where Q_i is the heat transfer rate from surface i; σ is the Stefan–Boltzmann
constant $(5.67 \times 10^{-8}$ kg s^{-3} K^{-4}); ε_i is the effective emissivity of surface i; F_{ij} is
the view factor from surface i to surface j which is a function of the area of surface
i and j. This view factor describes the fraction of thermal energy leaving the
surface of object i and reaching the surface of object j. In other words, F_{ij} is the
fraction of object j visible from the surface of object i. This quantity is also known
as the Radiation Shape Factor. Its unit is dimensionless. A_i is the area of surface i,
and T_i, T_j are absolute temperatures at surfaces i and j, respectively.

Having discussed the thermal analysis, let us now discuss the stress analysis.

3.6.2 Stress Analysis

3.6.2.1 Element Selection for Stress Analysis

Elements PLANE42 and SOLID45 are used to analyze stress field for 2D and 3D
analysis, respectively (Fig. 3.10). PLANE42 is used for 2D modeling of solid
structures. The element can be used either as a plane element (plane stress or plane
strain) or as an axisymmetric element. The element is defined by four nodes having
two DOF at each node, namely the translations in the nodal x and y directions.
SOLID45 is used for the 3D modeling of solid structures. The element is defined
by eight nodes having three DOF at each node, namely the translations in the nodal

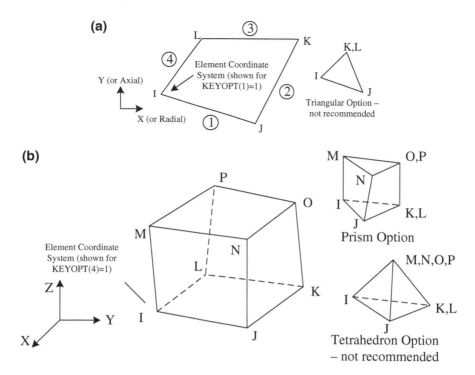

Fig. 3.10 a PLANE42 for 2D and **b** SOLID45 for 3D stress field analysis [18]

x, y, and z directions. Both PLANE42 and SOLID45 have plasticity, creep, swelling, stress stiffening, large deflection, and large strain capabilities.

3.6.2.2 Theoretical Aspect of Stress Analysis

Stress–strain relationships [18]. For linear elastic materials, the stress is related to the strain by Timoshenko [21]:

$$\{\sigma\} = [D]\{\varepsilon^{el}\} \tag{3.59}$$

where $\{\sigma\}$ is the stress vector $= [\sigma_{xx} \quad \sigma_{yy} \quad \sigma_{zz} \quad \tau_{xy} \quad \tau_{yz} \quad \tau_{xz}]^{T}$. Figure 3.11 gives the schematic drawing of the stress components in the finite element calculation. $[D]$ is the elastic matrix which is related to the Young's and shear modulus and Poisson's ratio. The total strain vector is $\{\varepsilon^{to}\}$ and can be expressed as $\{\varepsilon^{to}\} = [\varepsilon_{xx}^{to} \quad \varepsilon_{yy}^{to} \quad \varepsilon_{zz}^{to} \quad \varepsilon_{xy}^{to} \quad \varepsilon_{yz}^{to} \quad \varepsilon_{xz}^{to}]^{T}$. It is the sum of the elastic strain $\{\varepsilon^{el}\}$ and thermal strain $\{\varepsilon^{th}\}$. The thermal strain vector $\{\varepsilon^{th}\} = \Delta T [\beta_x \quad \beta_y \quad \beta_z \quad \beta_{xy}$ $\beta_{yz}\beta_{xz}]^{T}$, β is the thermal expansion coefficient, and $\Delta T = T - T_{ref}$. T is the temperature at the point of interest and T_{ref} is the reference temperature where the system is stress/strain free.

Equation 3.59 may also be inverted to:

$$\{\varepsilon^{to}\} = \{\varepsilon^{th}\} + [D]^{-1}\{\xi\} \tag{3.60}$$

And the "column normalized" format of $[D]^{-1}$ is:

$$[D]^{-1} = \begin{bmatrix} 1/E_x & -v_{xy}/E_y & -v_{xz}/E_z & 0 & 0 & 0 \\ -v_{yx}/E_x & 1/E_y & -v_{yz}/E_z & 0 & 0 & 0 \\ -v_{zx}/E_x & -v_{zy}/E_y & 1/E_z & 0 & 0 & 0 \\ 0 & 0 & 0 & 1/G_{xy} & 0 & 0 \\ 0 & 0 & 0 & 0 & 1/G_{yz} & 0 \\ 0 & 0 & 0 & 0 & 0 & 1/G_{xz} \end{bmatrix}$$

where E_x is Young's modulus in the x direction, v_{xy} is Poisson's ratio, and G_{xy} is shear modulus in the xy plane.

Strain and stress evaluation. The strains may be related to the nodal displacements by:

$$\{\varepsilon\} = [B]\{u\} \tag{3.61}$$

or

$$\{\varepsilon^{el}\} = [B]\{u\} - \{\varepsilon^{th}\}. \tag{3.62}$$

Here, $[B]$ is the strain-displacement matrix. $\{u\}$ is the nodal displacement vector given as $\{u_x \quad u_y \quad u_z\}^T$, where the x, y, z coordinates are defined in Fig. 3.11. A 2D example is given below to elaborate Eq. 3.61 in the case of plane stress:

Fig. 3.11 Schematic of the stress components in the finite element calculation

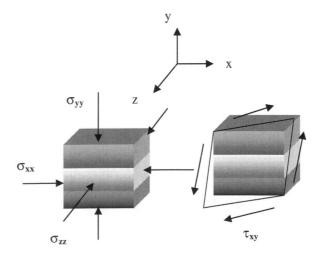

$$\{\varepsilon\} = \{ \varepsilon_x \quad \varepsilon_y \quad \varepsilon_{xy} \}^T \tag{3.63}$$

$$[B] = \begin{bmatrix} \frac{\partial N_1}{\partial x} & 0 & \frac{\partial N_2}{\partial x} & 0 & \cdots & \frac{\partial N_n}{\partial x} & 0 \\ 0 & \frac{\partial N_1}{\partial y} & 0 & \frac{\partial N_2}{\partial y} & \cdots & 0 & \frac{\partial N_n}{\partial y} \\ \frac{\partial N_1}{\partial y} & \frac{\partial N_1}{\partial x} & \frac{\partial N_2}{\partial y} & \frac{\partial N_2}{\partial x} & \cdots & \frac{\partial N_n}{\partial y} & \frac{\partial N_n}{\partial x} \end{bmatrix} \tag{3.64}$$

$$\{u\} = \{ u_{x1} \, u_{y1} \, \ldots \, u_{xn} \, u_{yn} \}^T \tag{3.65}$$

N_i is the element shape function as defined in Eq. 3.42.

The principal strains ε_0 can be calculated from the strain components by the cubic equation after the normal components (ε_x, ε_y, ε_z, ε_{xy}, ε_{xz}, ε_{yz}) as obtained above [22]:

$$\begin{vmatrix} \varepsilon_x - \varepsilon_0 & \frac{\varepsilon_{xy}}{2} & \frac{\varepsilon_{xz}}{2} \\ \frac{\varepsilon_{xy}}{2} & \varepsilon_y - \varepsilon_0 & \frac{\varepsilon_{yz}}{2} \\ \frac{\varepsilon_{xz}}{2} & \frac{\varepsilon_{yz}}{2} & \varepsilon_z - \varepsilon_0 \end{vmatrix} = 0 \tag{3.66}$$

Similarly, the principal stresses σ_0 can be calculated from the stress components using the cubic equation [22]

$$\begin{vmatrix} \sigma_x - \sigma_0 & \sigma_{xy} & \sigma_{xz} \\ \sigma_{xy} & \sigma_y - \sigma_0 & \sigma_{yz} \\ \sigma_{xz} & \sigma_{yz} & \sigma_z - \sigma_0 \end{vmatrix} = 0 \tag{3.67}$$

In the FEM for interconnect reliability, as current will be flowing through the interconnect, electrical analysis is also important, and will be discussed next.

3.6.3 Electrical Analysis

3.6.3.1 Element Selection for Electrical Analysis

This analysis type is used to determine the electrical potential in a conducting body created by an external application of voltage or current loads. From the solution, other parameters such as the conduction currents, electric field, and joule heating can be computed. An electrical analysis computes Joule heating from the electrical resistance and current in the conductor. This joule heating may be passed as a load to a subsequent thermal analysis using an imported load if the electrical analysis solution data is to be transferred to the thermal analysis. Similarly, an electrical analysis can accept a thermal condition from a thermal analysis to specify the temperatures in the body for material properties evaluation when they are temperature-dependent.

PLANE67 and SOLID69 are used to perform electrical analysis for 2D and 3D models, respectively (Fig. 3.12). PLANE67 has thermal and electrical conduction capability. Joule heating generated by the current flow is also included in the heat

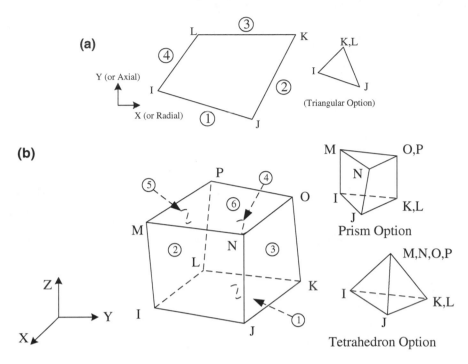

Fig. 3.12 **a** PLANE67 for 2D and **b** SOLID69 for 3D electrical analysis [18]

balance equation. The element has four nodes with two DOF, namely temperature and voltage, at each node. SOLID69 has a 3D thermal and electrical conduction capability. Joule heat generated by the current flow is also included in the heat balance equation. The element has eight nodes with two DOF, temperature and voltage, at each node.

3.6.3.2 Theoretical Aspect of Electrical Analysis

Electrical scalar potential. The electrical analysis is governed by the Maxwell's equations. By neglecting the time derivative of magnetic flux density, the system of Maxwell's equations reduces to [18]

$$\nabla \times \{H\} = \{J\} + \left\{\frac{\partial D}{\partial t}\right\} \tag{3.68}$$

$$\nabla \times \{E\} = \{0\} \tag{3.69}$$

$$\nabla \cdot \{B\} = 0 \tag{3.70}$$

$$\nabla \cdot \{D\} = \rho \tag{3.71}$$

As follows from Eq. 3.69, the electric field $\{E\}$ is irrotational, and can be derived from

$$\{E\} = -\nabla V \tag{3.72}$$

Here H is magnetic field, E is electrical field, J is current density, D is electrical displacement, B is magnetic flux density, ρ is resistivity of the material, and V is electrical potential.

The constitutive equations for the electric fields are [18]

$$\{J\} = [\sigma]\{E\} \tag{3.73}$$

$$\{D\} = [\varepsilon]\{E\} \tag{3.74}$$

where:

$$[\sigma] = \begin{pmatrix} \frac{1}{\rho_{xx}} & 0 & 0 \\ 0 & \frac{1}{\rho_{yy}} & 0 \\ 0 & 0 & \frac{1}{\rho_{zz}} \end{pmatrix} = \text{electrical conductivity matrix} \tag{3.75}$$

$$[\varepsilon] = \begin{pmatrix} \varepsilon_{xx} & 0 & 0 \\ 0 & \varepsilon_{yy} & 0 \\ 0 & 0 & \varepsilon_{zz} \end{pmatrix} = \text{permittivity matrix} \tag{3.76}$$

3.6.4 Coupled-Field Analysis

As mentioned earlier, coupled-field analysis is often needed in the FEM for a realistic problem. Coupled-field analysis takes into account the interaction between two or more fields. One can couple the two fields by applying results from the first analysis (e.g., thermal analysis) as load for the second analysis (e.g., stress analysis), as shown in Fig. 3.5. This is so-called the sequential coupled-field thermal stress analysis, where PLANE55 for thermal analysis should be used in connection with PLANE42 for stress analysis in 2D case. In 3D case, SOLID70 for thermal analysis is coupled with SOLID45 for stress analysis. The coupled-field analysis can further be extended by including the electrical field analysis. An example is the inclusion of Joule heating.

Joule heating is common when there is a current passing through an interconnect. The Joule heating contributes to the heat transfer process in an interconnect system, and thus changes the temperature and stress distributions. Therefore, with Joule heating, the problem is an electrical–thermal-stress coupling. In ANSYS, Joule heating is computed by elements with non-zero resistivity and non-zero current density applied. Elements PLANE13 and SOLID5 (Fig. 3.13) holds the capability to do this kind of electrical–thermal-stress coupled analysis for 2D and 3D, respectively.

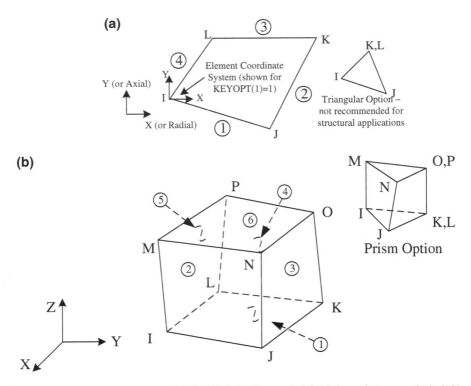

Fig. 3.13 a PLANE13 for 2D and **b** SOLID5 for 3D coupled-field thermal stress analysis [18]

Joule heating in element j for a static or transient electrical/magnetic analysis is computed as follows [18]

$$Q^j = \frac{1}{n} \sum_{i=1}^{n} [\lambda]\{J_i\} \cdot \{J_i\} \tag{3.77}$$

where n is the number of nodes in an element. $[\lambda]$ is the resistivity matrix, and $\{J_i\}$ is the current density in the element j at node i.

Besides sequential coupled-field analysis, steady-state thermal-electric conduction analysis allows for a simultaneous solution of thermal and electric fields. This coupled-field capability models joule heating for resistive materials and contact electric conductance as well as Seebeck, Peltier, and Thomson effects for thermoelectricity. For a completed electro-thermal and thermo-mechanical analysis, a finite element model is constructed using SOLID69 with the temperature and current (voltage) as input. Its thermal analysis result can be transferred as new input information for SOLID45 modeling for the subsequent thermo-mechanical analysis.

ANSYS also provides another option to solve multiphysics problem which is called the direct coupled-field analysis. In this method, only a single element will be used for both thermal and stress field analysis. Elements PLANE13 and SOLID5 are used to analyze the coupled-field thermal stress for 2D and 3D, respectively (Fig. 3.13). PLANE13 has a 2D magnetic, thermal, electrical, piezoelectric, and structural field capability with coupling between the fields. PLANE13 is defined by four nodes with up to four DOF per node. One option of the DOF combination is: UX (displacement in X direction), UY (displacement in Y direction), and TEMP (temperature), which is applicable for the thermal stress analysis. When used in a structural analysis, PLANE13 has large deflection, large strain, and stress stiffening capabilities. SOLID5 has a 3D magnetic, thermal, electric, piezoelectric, and structural field capability with limited coupling between the fields. The element has eight nodes with up to six DOF at each node. One optional DOF combination suitable for the thermal stress analysis is: UX (displacement in X direction), UY (displacement in Y direction), UZ (displacement in Z direction), and TEMP (temperature).

3.7 ANSYS Parametric Design Language (APDL)

ANSYS Parametric Design Language (APDL) [23] is a powerful scripting language provided by ANSYS through which you can build your model in terms of parameters and variables. Under APDL environment, all ANSYS commands can be used. In addition, it encompasses a wide range of other features such as repeating a command, macros, if then else branching, do-loops, scalar, vector, and matrix operations. Users familiar with basic programming languages will find the APDL more convenient to use than the conventional GUI mode, especially when one needs to input their own equations in ANSYS. The user-defined program, which is sometimes known as subroutine, can be imported into ANSYS, and then de-coded automatically in the ANSYS environment.

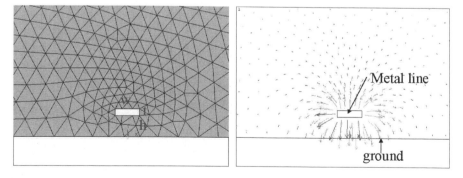

Fig. 3.14 A 2D finite element mesh and the electrical field vector of a metal above a ground plane for verification purpose

Two simple examples are given below showing how to calculate the electric field and capacitance for a case with a metal above a ground plane as shown in Fig. 3.14.

Example I This example will provide a solution to extract the electrical field for the results shown in Fig. 3.14.

```
/clear      ! Clears the database

/prep7     ! Enters the model creation pre-processor

!The following are Variable Definition Using the "=" Command.

h=1        ! To define distance between metal and the ground.

w=2.72    ! To define metal width.

t=0.802   ! To define metal thickness.

xt=100+10*w   !To define the dimension in x-direction for the domain under study

yt=h+t+10    !To define the dimension in y-direction for the domain under study

v1=2       !To define the voltage

v2=0       !To define the ground


et,1,121   !To define element type.

emunit,epzro,1    !To specify units for magnetic field.

mp,perx,1,1    !To define material property


rectng,-xt/2,xt/2,0,yt   !To create a rectangular area representing the whole domain

rectng,-w/2,w/2,h,h+t   !To create a rectangular area representing the metal

aovlap,all              !To overlap the above two areas


numcmp,area     !To compress the numbering of the above areas

smrtsiz,1           !To specify meshing parameters for element sizing

mshape,1        !To specify the element shape for meshing

amesh,2         !To do area meshing


!To select those nodes for applying voltage "v1"

lsel,s,,,5,8

nsll,s,1

d,all,volt,v1

!To select those nodes for applying "v2"

lsel,s,,,1

nsll,s,1

d,all,volt,v2

allsel,all

finish


/solu      !To enter the Solution Processor

solve     !To solve the problem

finish
```

Example II This example uses the same geometry as Example I, and provides solution to calculate the capacitance for the structure shown in Fig. 3.14. The capacitance with various dimensions is then given in Table 3.6.

```
/clear     ! Clears the database

/prep7     ! Enters the model creation pre-processor

!The following are Variable Definition Using the "=" Command.
h=1        ! To define distance between metal and the ground.
w=2.72   ! To define metal width.
t=0.802   ! To define metal thickness.
xt=100+10*w   !To define the dimension in x-direction for the domain under study
yt=h+t+10     !To define the dimension in y-direction for the domain under study

et,1,121   !To define element type.
emunit,epzro,1   !To specify units for magnetic field.
mp,perx,1,1      !To define material property

rectng,-xt/2,xt/2,0,yt   !To create a rectangular area representing the whole domain
rectng,-w/2,w/2,h,h+t   !To create a rectangular area representing the metal
aovlap,all               !To overlap the above two areas

numcmp,area    !To compress the numbering of the above areas
smrtsiz,1            !To specify meshing parameters for element sizing
mshape,1      !To specify the element shape for meshing
amesh,2       !To do area meshing

!To select those nodes for defining component "cond1"
lsel,s,,,5,8
nsll,s,1
cm,cond1,node   !To group the selected nodes into a component named "cond1"
!To select those nodes for for defining component "cond2"
lsel,s,,,1
nsll,s,1
cm,cond2,node      !To group the selected nodes into a component named "cond2"
allsel,all
finish

/solu    !To enter the Solution Processor
!To calculate the capacitance between components "cond1" and "cond2"
cmatrix,1,'cond',2,0
finish
```

Table 3.6 The capacitance of the structure given in Fig. 3.14, calculated from FEM (ANSYS)

w/h	t/h	$C2\ (=C/\varepsilon*1)$ Numerical
1.12	0.318	3.620
2.01	0.485	4.840
2.52	0.318	5.256
2.72	0.802	5.866
3.18	0.936	6.451
3.63	0.802	6.893
3.68	0.485	6.692
3.79	1.116	7.216
4.24	0.936	7.595
5.44	1.2	9.000
6.36	0.802	9.760
7.42	0.936	10.947
8.21	1.805	12.106
9.85	1.805	13.788
11.9	1.453	15.736
14.78	1.805	18.779
22.22	0.407	26.790
58.25	7.113	63.141

3.8 Conclusion

In this chapter, the history of FEM is introduced. The principle of FEM is elaborated through simple example. Several commercial software, such as the popular ANSYS and ABAQUS for FEM are described. The basic steps in using a commercial software to perform the FEA is discussed. Finally, the methodology of FEM for interconnect study in ULSI is given, including the thermal analysis, stress analysis, electrical analysis, and more importantly, the coupled-field analysis. The selection of element types for each analysis is also given.

References

1. Courant R (1943) Variational methods for the solution of problems in equilibrium and vibrations. Bull Am Math Soc 49:1–23
2. Turner MJ, Clough RW, Martin HC, Topp LJ (1956) Stiffness and deflection analysis of complex structures. J Aeronaut Sci 23:805–824
3. Noor AK (1991) Bibliography of books and monographs on finite element technology. Appl Mech Rev 44:307–317
4. Moes N, Cloirec M, Cartraud P, Remacle J-F (2003) A computational approach to handle complex microstructure geometries. Comput Methods Appl Mech Eng 192:3163–3177
5. Weide-Zaage K, Dalleaua D, Yu X (2003) Static and dynamic analysis of failure locations and void formation in interconnects due to various migration mechanisms. Mater Sci Semicond Process 6:85–92

6. Shen Y-L, Guo YL, Minor CA (2000) Voiding induced stress redistribution and its reliability implications in metal interconnects. Acta Mater 48:1667–1678
7. Nguyen HV, Salm C, Wenzel R, Mouthaan AJ, Kuper FG (2002) Simulation and experimental characterization of reservoir and via layout effects on electromigration lifetime. Microelectron Reliab 42:1421–1425
8. Zhang YW, Bower AF, Xia L, Shih CF (1999) Three dimensional finite element analysis of the evolution of voids and thin films by strain and electromigration induced surface diffusion. J Mech Phys Solids 47:173–199
9. Kuroda S, Kawai Y, Onoda H, Nishi K (1991) Viscoelasticity-based time-dependent stress model for a submicron aluminum interconnect. IEDM, pp 713–716
10. Wallace B, Lee Y-H, Pantuso D, Wu K, Mielke N (1999) Thermo-mechanical stress-induced voiding in a tungsten-AlCu interconnect system. In: 37th annual international reliability physics symposium, pp 303–309
11. Im S, Banejee K, Goodson KE (2002) Modeling and analysis of via hot spots and implications for ULSI interconnect reliability. In: 40th annual international reliability physics symposium, pp 336–345
12. Huang TC, Yao CH, Wan WK, Hsia CC, Liang MS (2003) Numerical modeling and characterization of the stress migration behavior upon various 90 nanometer Cu/low-k interconnects. In: Proceedings of the IEEE international interconnect technology conference, pp 207–209
13. Chen F, Chanda K, Gill I, Angyal M, Demarest J, Sullivan T, Kontra R, Shinosky M, Li J, Economikos L, Hoinkis M, Lane S, McHerron D, Inohara M, Boettcher S, Dunn D, Fukasawa M, Zhang BC, Ida K, Ema T, Lembach G, Kumar K, Lin Y, Maynard H, Urata K, Bolom T, Inoue K, Smith J, Ishikawa Y, Naujok M, Ong P, Sakamoto A, Hunt D, Aitken J (2005) Investigation of CVD SiCOH low-k time-dependent dielectric breakdown at 65 nm node technology. In: Proceedings of the international reliability physics symposium, pp 501–507
14. Tsu R, McPherson JW, McKee WR (2000) Leakage and breakdown reliability issues associated with low-k dielectrics in a dual-damascene Cu process. In: IEEE 38th annual international reliability physics symposium, San Jose, California, pp 348–353
15. Zienkiewicz OC (1977) The finite element method, 3rd edn. McGraw-Hill, New York
16. Becker EB, Carey GF, Oden JT (1981) Finite elements: an introduction, vol I. Prentice-Hall, Englewood Cliffs
17. Moaveni S (1999) Finite element analysis: theory and application with ANSYS. Prentice Hall, Englewood Cliffs
18. Swanson Analysis System, Inc (1999) ANSYS theory reference, 5.6 edn. Swanson Analysis System, Inc, Houston
19. White FM (1988) Heat and mass transfer. Addison-Wesley, New York
20. Rohsenow WM, Hartnett JP, Ganic EJ (1985) Handbook of heat transfer fundamentals. McGraw-Hill, New York
21. Timoshenko SP, Goodier JN (1987), Theory of elasticity, 3rd edn. McGraw-Hill, New York
22. Dieter GE (1988) Mechanical metallurgy. McGraw-Hill, London
23. ANSYS, Inc (2009) ANSYS parametric design language guide, 12th edn. ANSYS, Inc., Canonsburg

Chapter 4
Finite Element Method
for Electromigration Study

4.1 Introduction

In Chap. 3, we introduce the basic concept and the general theory of finite element method (FEM) for electrical, thermal, mechanical, and coupled-field multi-physics analysis. EM is a complicated physical and material phenomenon that involves the analysis of electro-thermo-mechanical coupled-field analysis, and governed by various partial differential equations. FEM is able to solve the partial differential equation and handle complex geometries (and boundaries) with relative ease. Therefore, several EM studies employ the FEM for more complete investigation on different interconnect structures. In this chapter, we will provide an overview on the application of FEM for EM study.

4.2 A Review on the Electromigration Modeling Using FEM

As discussed in Chap. 2, it is the presence of atomic flux divergence (AFD) that renders void and hillock formations in interconnections. Since the lifetime of a metal line is closely related to the void and hillock formation in the interconnect [1], the values of AFD determines the rate of void and hillock growth rate, and the EM lifetime is inversely proportional to the AFD as will be elaborated further in this chapter. Also, the maximum AFD locations will be the void and hillock locations in a metal line.

The presence of the surrounding materials, such as the barrier metals and the oxide or low K dielectric will cause nonuniform temperature/stress distributions along an interconnect. This nonuniformity enhances the disparity of the atomic diffusivity along a diffusion path which in turn increases the AFD. Hence, the distributions of the current density, temperature and hydrostatic stress in a metal line are key affecting factors on the value of AFD [2]. In 2D or 3D EM simulation,

C. M. Tan et al., *Applications of Finite Element Methods for Reliability Studies on ULSI Interconnections*, Springer Series in Reliability Engineering, DOI: 10.1007/978-0-85729-310-7_4, © Springer-Verlag London Limited 2011

FEM is often used to obtain the basic physical quantities such as current density, temperature and stress. AFD formulation and calculation are then performed based on the result of the Finite element analysis (FEA). Several EM models are developed based on AFD, but their mathematical implementation and theoretical derivation are different. In this section, we will provide a review on EM simulation methodologies reported in the literature.

4.2.1 Computation Methods for Atomic Flux Divergence

Sasagawa et al. reported their detailed study on AFD in 1998 [3, 4]. They proposed a calculation method of AFD for both unpassivated polycrystalline line and bamboo line. Later on, the calculation method was also applied to passivated polycrystalline line including the effect of the back flow stress in their subsequent work [5]. In their study, the AFD was identified as a parameter governing the void formation, and the distributions of the current density and temperature as well as the material properties of the metal thin film were considered in the computation of the AFD. Lifetime and failure site in a passivated/unpassivated metal line were predicted by means of the numerical simulation of the processes of the void initiation and void growth based on the AFD, and the change in the distributions of the current density and temperature with void growth was taken into account.

For the case of unpassivated metal lines [4], they adopted the basic atomic flux formula as follows [6]:

$$|J| = \frac{ND_0}{k_B T} \exp\left(-\frac{E_{a,gb}}{k_B T}\right) Z^* e \rho j^*. \tag{4.1}$$

where J is the atomic flux vector along the direction of the dominant grain boundary, N denotes the atomic density, D_0 is a prefactor for diffusion, k_B is Boltzmann's constant, T is absolute temperature, $E_{a,gb}$ is activation energy for grain boundary diffusion, Z^* is effective valence, e is electronic charge, ρ denotes the temperature-dependent resistivity expressed as $\rho = \rho_0\{1 + \alpha(T - T_s)\}$, ρ_0 and α are the resistivity and its temperature coefficient at the substrate temperature T_s, respectively. j^* is a component of the current density vector in the direction of J.

For the case of passivated metal line [5], they adopted the basic atomic flux formula as [7]

$$|J| = \frac{ND_0}{k_B T} \exp\left(-\frac{E_{a,gb} - \sigma\Omega}{k_B T}\right)\left(Z^* e \rho j^* + \Omega\frac{\partial\sigma}{\partial l}\right). \tag{4.2}$$

The activation energy is adjusted and additional stress-related terms are added. Here, σ is the hydrostatic stress, Ω is atomic volume, and $\frac{\partial\sigma}{\partial l}$ is component of the stress gradient in the direction of J.

In 1999, Rzepka et al. presented their 3D finite element simulator of EM in interconnect metal lines [8]. Their model simulated the migration of matter due to various driving forces, namely the concentration gradients, the mechanical stress gradient, the electrical field, the temperature gradients, and the surface tension due to the changes in surface curvature. Their work provided the pioneering study on the various driving forces of EM through a 3D simulation.

In their work, they creatively used the thermal analysis routine to model the diffusion phenomena. The basic mechanism that allows matter to move in solid is diffusion with the following fundamental relations

$$J = -D\nabla c, \tag{4.3}$$

$$\frac{\partial c}{\partial t} = -\nabla J, \tag{4.4}$$

where J, D, c, and t denote the diffusion flux, the diffusion coefficient, the concentration, and the time respectively. They found that these fundamental formulae are analogous to the heat flux equation in thermal analysis which was found by Adolf Fick in 1855 [9]. Because of this similarity, the atomic concentration change and atomic flux due to the various driving forces can be generalized and expressed as follows:

$$\frac{\partial c}{\partial t} = -\sum_i \nabla J_i = -\sum_i \nabla\left[-\frac{Dc_0}{k_B T}\nabla(V_i)\right] = \sum_i \nabla(Mc_0 F_i), \tag{4.5}$$

where c_0 is the initial concentration, M is mobility of that particular matter, and J_i is individual diffusion flux (thus ∇J_i is AFD) that results from the driving force F_i, which itself is caused by the gradient of the corresponding potential, V_i. The V_i for the atomic diffusion in EM are the following:

$$
\begin{aligned}
\text{Concentration}, \text{c}_c &= \Omega k_B T c \\
\text{Mechanical stress}, \text{s}_\sigma &= -\Omega\sigma \\
\text{Electrical potential}, \phi \quad V_\phi &= Z^* e\phi \\
\text{Temperature}, T \quad V_T &= Q^* \ln T \\
\text{Surface curvature}, \kappa \quad V_\kappa &= \gamma\Omega\kappa
\end{aligned} \tag{4.6}
$$

By applying the thermal routine of ANSYS multiply, the individual fluxes and AFD can be sequentially calculated, thus the change in the total atomic concentration can be obtained.

In 2001, Dalleau and Weide-Zaage reported their study on 3D voids simulation in chip metallization structures [10]. In their study, the different atomic flux due to different driving forces such as the electron wind force (EWF) migration ($\vec{J}_{EWM}$), the thermo-migration ($\vec{J}_{th}$) due to temperature gradient-induce driving force (TGIDF), and the stress migration ($\vec{J}_{str}$) due to stress gradient-induced driving force (SGIDF) were modeled respectively as follows:

$$\vec{J}_{EWM} = \frac{N}{k_B T}eZ^*\vec{j}\rho D_0 \exp\left(-\frac{E_A}{k_B T}\right), \tag{4.7}$$

$$\vec{J}_{\text{th}} = -\frac{NQ^*D_0}{k_BT}\exp\left(-\frac{E_A}{k_BT}\right)\nabla T, \tag{4.8}$$

$$\vec{J}_{\text{str}} = \frac{N\Omega D_0}{k_BT}\exp\left(-\frac{E_A}{k_BT}\right)\nabla \sigma_H, \tag{4.9}$$

where E_a is the measured activation energy, Q^* is the heat of transport, and σ_H is the local hydrostatic stress value ($\sigma_H = (\sigma_{xx} + \sigma_{yy} + \sigma_{z)z}/3$). Based on the above atomic flux equations, they derived the approximated AFD for the respective driving forces as follows:

$$\text{div}(\vec{J}_{\text{EWM}}) = \left(\frac{E_A}{k_BT} - \frac{1}{T} + \alpha\frac{\rho_0}{\rho}\right) \cdot \vec{J}_{\text{EWM}} \cdot \nabla T, \tag{4.10}$$

$$\text{div}(\vec{J}_{\text{th}}) = \left(\frac{E_A}{k_BT} - \frac{3}{T} + \alpha\frac{\rho_0}{\rho}\right) \cdot \vec{J}_{\text{th}} \cdot \nabla T + \frac{NQ^*D_0}{3k_B^3T^3}j^2\rho^2e^2\exp\left(-\frac{E_A}{k_BT}\right), \tag{4.11}$$

$$\text{div}\left(\vec{J}_{\text{str}}\right) = \left(\frac{E_A}{k_BT} - \frac{1}{T}\right)\vec{J}_{\text{str}}\nabla T + \frac{2EN\Omega D_0\alpha_l}{3(1-v)k_BT}\exp\left(-\frac{E_A}{k_BT}\right)\left(\frac{1}{T} - \alpha\frac{\rho_0}{\rho}\right)\nabla T^2$$
$$+ \frac{2EN\Omega D_0\alpha_l}{3(1-v)k_BT}\exp\left(-\frac{E_A}{k_BT}\right)\frac{j^2\rho^2e^2}{3k_B^2T} \tag{4.12}$$

The coupled-field analyses, namely electrical–thermal analysis and thermal–mechanical analysis were then performed using ANSYS. The electrical–thermal element *Solid 69* is used for the electrical–thermal analysis, and the result of the analysis is then used as an input load for the thermo-mechanical analysis using coupled-field analysis element *Solid 45*. Based on their analysis results, the distributions of the current density, temperature, hydrostatic stress and their respective gradients are obtained; and the values of AFD due to these three driving forces, respectively, were calculated according to Eqs. 4.10–4.12. Both the failure sites of the test structure and the lifetime of the test structure were predicted by their method.

In their subsequent work, the formula was also applied to different test structures, such as meander structure, pad structure [10], and dual-damascene structure [11] for both the Al and Cu technologies [12]. Adopting the similar AFD formula, Tan et al. studied the intrinsic EM performance of deep submicron Cu and Al interconnects, including the effect of surface migration and void nucleation [2]. In their subsequent work, the method was also successfully used to explain many experimental observations [13, 14].

However, the AFD formulation according to Eqs. 4.10–4.12 is no longer accurate in predicting void nucleation site as interconnects continue to scale down [15] due to the invalidity of the assumptions made in the derivations of Eqs. 4.10–4.12, as line becomes narrower. A revised AFD formulation based on the Green's theorem was then proposed by Tan et al. [15], and its validity was verified through the study of the reservoir effect in EM. The elaboration on this point will be found in Sect. 4.4 in detail.

4.2.2 Voiding Mechanism Simulation

With the knowledge of AFD, the mass transport through a given controlled volume at any location in a test structure can be obtained, and thus the process of voiding can be simulated. In the work of Sasagawa [4, 5], the metal line was divided into elements of smaller size to give more realistic results. The thickness of the element was changed by the procedure as shown in Fig. 4.1. Through this procedure, the growth of voids in the metal line was simulated.

Based on the FEM and the AFD formula, the distribution of AFD in each element can be calculated. Each calculation step was assigned to a realistic time. The volume decrement in each element per calculation step in the simulation was calculated based on the volume of the element, the realistic time corresponding to one calculation step, the atomic volume, and the value of AFD. The current density and temperature in the metal line were computed again according to the new thickness of each element using FEM after each iteration. The decrement of the thickness of the element was regarded as void that was being formed. Following this procedure, the 2D simulator was able to simulate the growth of the voids and predict the failure time when the void became fatal to the interconnects.

However, in the work of Dalleau and Weide-Zaag [10–12] as well as Tan [2], the process of void growth was simulated differently as shown in Fig. 4.2. The elements in the meshed structure were physically deleted to simulate the void growth. In Dalleau's study, the deletion was realized by changing the corresponding material properties to negligible electrical and thermal conductivities, as well as a negligible Young's modulus value for stress analysis. In Tan's study, those selected elements which have the highest AFD value was deleted from the

Fig. 4.1 Simulation flow chart in Sasagawa's study

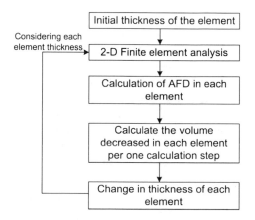

Fig. 4.2 The simulation flow
chart adopted by Dalleau
et al. [10] and Tan et al. [15]

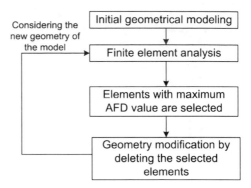

finite element model. In both simulations, the structure was automatically modified, and the FEA was repeated until the failure condition was reached. The main
challenge of void growth simulation was to take the time factor into consideration.

Although all the three void growth simulation methodologies have good
accuracy to simulate the void growth location and the void shape evolution, the
attempt to integrate the time factor in the simulation was not so successful. The
lifetime prediction for each simulation methodology will be discussed in detail in
the next section.

4.2.3 Lifetime Prediction

The prediction of the lifetime of a test structure depends on the method used to
determine the realistic time in the simulation and the specified condition/criterion
beyond which the elements do not exist anymore. In the work of Sasagawa et al.
[4, 5], the calculation process of the numerical simulation for the lifetime prediction was carried out repeatedly until the metal line fails where the entire line
width was either occupied by the elements whose temperature exceeds the melting
point of the metal due to Joule heating and/or the thickness of the elements become
smaller than a pre-defined infinitesimal threshold value. Based on their experimental measurement and the hypothesis of the effective width of the slit void, they
determined that the threshold value of the effective width should be 2×10^{-3}
times the initial film thickness. By assigning one calculation step to a "realistic
time," the lifetime of the test structure was predicted. Although the method of
prediction was found to have a good agreement with their experimental observations, the assignment of the "realistic time" to the calculation step was not
rigorous theoretically, and its extension to general EM test would be difficult.

Different prediction approaches were adopted in Dalleau et al.'s study [10–12].
The time-dependent evolution of the local atomic concentration was expressed as

$$\text{div}(J_{\text{EWM}} + J_{\text{th}} + J_{\text{str}}) + \frac{\partial N}{\partial t} = 0 \qquad (4.13)$$

$$\text{div}(J_{\text{EWM}} + J_{\text{th}} + J_{\text{str}}) = \text{div}(J_{\text{total}}) = N \cdot f(j, T, \sigma_{\text{H}}, E_{\text{a}}, D_0) \tag{4.14}$$

where f is a function including various physical parameters. Combining Eqs. 4.13 and 4.14, Eq. 4.13 can be rewritten as

$$Nf + \frac{\partial N}{\partial t} = 0. \tag{4.15}$$

Solving Eq. 4.15, yields the theoretical evolution of the atomic density as

$$N = N_0 e^{-ft} \tag{4.16}$$

and N_0 is the initial value of the atomic concentration. In other words,

$$t = \frac{1}{f} \ln \frac{N_0}{N}. \tag{4.17}$$

In their dynamic simulation, the element was considered to be a void when the atomic concentration reaches 10% of the initial concentration ($N/N_0 = 10\%$). The TTF was then the summation of the time needed to delete all the relevant elements for the creation of a critical void formation. Therefore, TTF in Eq. 4.17 can be rewritten as

$$TTF = \sum_{i=1}^{n} t_i = \ln 10 \sum_{i=1}^{n} \frac{1}{f} \tag{4.18}$$

where n is the number of deleted elements simulating the entire void formation to reach a specified failure criteria.

However, there are several limitations for the TTF calculation using Eq. 4.18 which are listed as follows:

1. TTF computed using Eq. 4.18 is the time for void-growing process. For nucleation-limited failure, void nucleation time is much longer than the void-growing time, and the void nucleation time is not accounted for in Eq. 4.18.
2. The element is considered to be a void with 10% reduction in the atomic concentration. The actual time needed for the element to be actually empty can be much larger [16] and it must be determined for accurate TTF estimation.

4.2.4 Summary

The various 3D FEM simulations of EM are discussed. The EM kinetic due to various driving forces, such as EWF temperature gradients, stress gradients, and concentration gradients, are quantitatively modeled through the calculation of their corresponding AFDs. Rather than solving the EM equation in a multi-dimension model, the required physical parameters are at first calculated using FEM.

The AFD due to different EM-driving forces is then calculated locally. With the knowledge of AFD at every node in the FEM model, the formation and evolution of the void can be simulated. Using this simulation methodology, the EM kinetic can be evaluated not only in the voiding region, but also within the entire EM test structure.

Due to the capability of the simulation in 3D structure, the effect of surrounding material, the shape of the metal line, or the metal stack structure in an interconnect system can be simulated in a single model. Furthermore, the EM simulator using FEM is capable of studying the EM kinetics locally around the region of voiding, and addressing the EM performance and potential reliability weakness of a metal interconnect from a microscopic point of view.

In most of the AFD-based EM simulation, the formulation and calculation of AFD is based on the approximated derivative of the atomic fluxes [4, 5, 10–12]. While the approximated AFD formulae have been successful in explaining many EM experimental observations, the assumptions adopted to simplify the derivation of AFD from atomic flux may not be valid in nano-interconnects as will be elaborated in Sect. 4.4. In Sect. 4.4, we will discuss in more detail, the risk of adopting the approximated AFD formulas for the interconnects in nanoscale and the improved methodology for AFD formulation using the Green function. Also, some of the EM simulations rely on the commercial FEM software to calculate the AFD from the atomic flux. However, due to the limitations of the post-processing module of commercial FEA software, a complicated mathematical operation is limited. In the following section, we introduce a methodology, leveraging on the advantage of both commercial FEA software and commercial mathematical software to calculate the AFD efficiently and accurately.

4.3 Enhanced Finite Element Method Through Matlab

In this section, we will discuss a newly developed simulation methodology by combining the commercial software ANSYS and Matlab reported by Li [17], so that the computational limitation of ANSYS can be complemented by Matlab, and AFD can thus be calculated directly. We will present an example of EM modeling on ULSI Cu interconnect using this extended capability of ANSYS static thermal analysis algorithm with Matlab linkage, and taking into account the effect of microstructure inhomogeneities of Cu thin film, surrounding materials of the test structure, and the mismatch of their coefficients of thermal expansion.

4.3.1 Finite Element Model

The finite element model of a Cu thin film with its surrounding materials is created using ANSYS Multi-physics as shown in Fig. 4.3. The barrier layer is

Fig. 4.3 Finite element
model of Cu thin film and its
surrounding materials.
Reprinted from Li and Tan
[17], copyright © 2007 with
permission from Elsevier

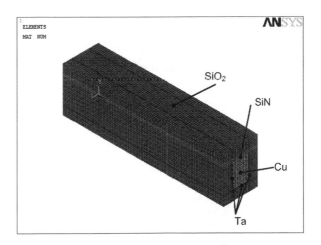

Table 4.1 Material properties used in finite element model (Cu interconnect)

Material	Young's modulus (GPa)	Poisson ratio	Yield stress (MPa)	Thermal conductivity (W/mK)	Coefficient of thermal expansion (/°C)	References
Cu	129.8	0.339	676 (20°C) 165 (350°C)	379	16.5×10^{-6}	[2, 18]
Ta	186.2	0.35	–	53.65	6.48×10^{-6}	[19]
SiN	265	0.27	–	0.8	1.5×10^{-6}	[19]
SiO$_2$	71.4	0.16	–	1.75	0.68×10^{-6}	[19]

Ta, the capping layer is SiN, and the entire metallization stack is embedded in
SiO$_2$.

In the simulation, SiO$_2$, SiN, and Ta were taken as isotropic linear elastic solids,
and Cu was characterized as an isotropic elastic-perfectly plastic solid [18]. Their
material properties are summarized in Table 4.1.

The Cu thin film has a height of 0.4 μm and a width of 0.25 μm. The thickness
of the SiN capping layer and the Ta barrier layer are 0.05 and 0.025 μm,
respectively. Figure 4.4 shows the hypothetical grain structure of the Cu thin film.
The arrow shows the direction of the electron flow.

The simulation involves complex electrical–thermal–mechanical interaction,
and two coupled-field analyses are employed in the study, namely the electrical–
thermal coupled-field analysis and thermal–mechanical coupled-field analysis.

In the procedure for generating the element, element type *Solid 69* with tetrahedral shape is used for the electrical–thermal analysis; an equivalent element type
Solid 45 was used for the thermal–mechanical analysis. The stress-free temperature in the model is set at 350°C which is the final annealing temperature of the
wafers. The simulated test conditions are 300°C and 3 mA/cm^2.

Fig. 4.4 Different grains in the finite element model (surrounding materials are not shown for the sake of clarity). Reprinted from Li and Tan [17], copyright © 2007 with permission from Elsevier

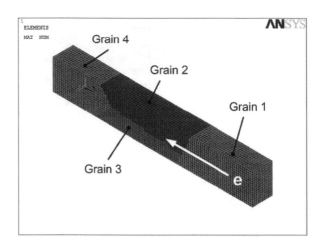

Table 4.2 Physical parameter adopted in the simulation [17]

Physical parameter	Symbol	Value (Cu)
Atom concentration	N	$8.44 \times 10^{10}/\mu m^3$
Effective valence charge	Z^*	4
Pre-exponential factor	D_0	$7.8 \times 10^7 \ \mu m^2/s$
Resistivity	ρ_0	$1.67 \times 10^{-14} \ T\Omega \ \mu m$
Temperature coefficient of resistivity	α	0.0039
Heat of transfer	Q^*	$2.773 \times 10^{-8} \ pJ$
Poisson ratio	ν	0.34
Young's modulus	E	$1.3 \times 10^5 \ MPa$
Coefficient of thermal expansion	α_l	$16.5 \times 10^{-6}/K$
Activation energy for lattice diffusivity	E_A	2.2 eV

4.3.2 Simulation Methodology

The physical environment of the Cu thin film during EM test was obtained through the aforementioned electrical–thermal coupled-field analysis and thermal–mechanical coupled-field analysis. The mathematical expressions of atomic fluxes due to EWF, TGIDF, and SGIDF used were according to Eqs. 4.7–4.9. The physical parameters used in the simulation are summarized in Table 4.2.

To examine the effect of the different driving forces of EM, their corresponding AFD needs are calculated. However, ANSYS post-processing module has the limitation to compute the divergence of a flux vector field obtained from its solution. A methodology to transfer data and system control between ANSYS and Matlab is developed, where the EM atomic fluxes are calculated in ANSYS as mentioned earlier and the atomic flux results are sent to Matlab for the computation of the AFD due to the various EM-driving forces. After that, the AFD result

Fig. 4.5 Flowchart of
simulation using ANSYS and
Matlab. Reprinted from Li
and Tan [17], copyright
© 2007 with permission from
Elsevier

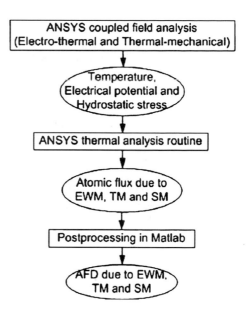

is transferred back to ANSYS for further analysis. The flowchart of the simulation
and mathematical implementation is illustrated in Fig. 4.5.

The transfer is performed automatically according to a pre-coded command
flow without user intervention. Using the ANSYS command "*VWRITE," the
value of the respective atomic flux is written into a file in a formatted sequence.
Please note that "*VWRITE" command must be contained in an externally pre-
pared file and read into ANSYS (i.e., *USE, /INPUT, etc.). After the data are
extracted from ANSYS and stored in a txt file, the string/SYS, C:\MAT-
LAB7\bin\win32\MATLAB.exe/r filename.m is executed. By executing the string,
the system control will be passed to Matlab environment without quitting ANSYS
and the filename.m is the Matlab m file used to extract the ANSYS data from the
txt file. In the m file, the Matlab command "load" and a few nested "for" loops
are used to read and sort the ANSYS data into Matlab from the txt file to calculate
the flux divergence in a 3D matrix. To quit Matlab, the command "exit" is executed, and the system control will then return to ANSYS and continue from the
ANSYS command just before the control was passed to Matlab. The data and
system control exchange method can be applied to other simulation involved FEM
using ANSYS to enhance the data processing capability.

4.3.3 Simulation and Experimental Result

In recent Cu interconnect technology, special treatment of the interface between
the capping layer and Cu thin film has successfully retarded the atomic diffusion
along the fast interface diffusion path by increasing the bonding strength of the

capping layer to Cu. With the absence of the fastest diffusion path, the second-fastest diffusion path, the grain boundaries become dominant [20–22] and this could introduce new EM failure mechanism.

The EM phenomenon of a Cu interconnect thin film is simulated assuming the interface diffusion in Cu interconnections is retarded by the processing technology improvement. The activation energies for diffusion along important mass transport pathways in the simulation are listed in Table 4.3.

By virtue of the assumption of interface improvement, the activation energy of the interface diffusion is chosen to be the maximum among the reported value (typical value is 0.9–1.2 eV [23]).

Figure 4.6 shows the plot of the AFD distribution in the Cu thin film using the new proposed ANSYS + Matlab simulation. The potential void nucleation location is found to be at the cap layer–grain boundary intersection due to its high AFD value.

The simulation result is consistent with the experimental observation as shown in Fig. 4.7. The Cu EM sample was fabricated with SiH_4 surface treatment at the capping layer interface. Failure analysis found that the EM-induced void indeed occurs at the cap layer–grain boundary intersection of Cu line after EM test, which agrees with the predicted void nucleation location in the simulation.

However, the observed failure location is not accurately predicted in the AFD plot if the approximated AFD formula is used as shown in Fig. 4.8. The simulation fails to predict the potential void initiation location in this circumstance.

Table 4.3 Activation energy used in Cu interconnect EM simulation [23]

Diffusion path	Activation energy, E_a (eV)
Lattice	2.1
Grain boundary	1.2
Interface	1.2

Fig. 4.6 Maximum AFD value in Cu thin film using enhanced ANSYS simulation. Reprinted from Li and Tan [17], copyright © 2007 with permission from Elsevier

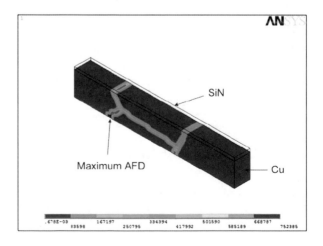

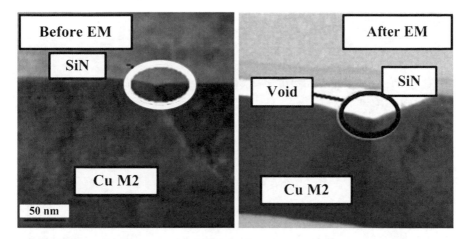

Fig. 4.7 Experimental observation of void at the intersection of grain boundary and cap layer in Cu interconnect after EM (courtesy from Prof Subodh Mhaisalkar, NTU MSE). Reprinted from Li and Tan [17], copyright © 2007 with permission from Elsevier

Fig. 4.8 Maximum AFD value in Cu thin film using approximated AFD formula. Reprinted from Li and Tan [17], copyright © 2007 with permission from Elsevier

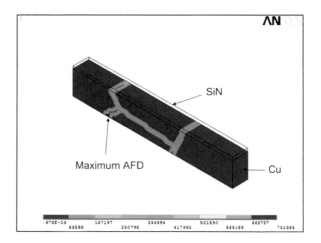

From the above example, we can see that the link between ANSYS and MatLab improves the efficiency of post-processing of the FEA for EM simulation. The methodology also avoids any assumptions adopted in the AFD formula derivation.

4.4   Improved EM Simulation Methodology Based on the Green Function

In the conventional formulation of AFD due to different EM-driving forces given by Eqs. 4.10–4.12, two assumptions were made in order to simplify the calculation. In this section, the justification of the assumptions made will be discussed in detail.

4.4.1 Derivation of Atomic Flux Divergence Using Green Function

In the derivations of Eqs. 4.11 and 4.12, the following energy balance equation in a thin film conductor at steady-state is used, with the third term assumed to be negligible due to its small contribution [24].

$$j^2\rho + K\nabla^2 T - \frac{k(T - T_s)}{\text{th}} = 0 \tag{4.19}$$

where K and k are the thermal conductivity of the metallic film and the dielectrics, respectively, T and T_s are the temperatures of the metallic film and the substrate, t is metallic film thickness, and h is dielectric film thickness. The three terms in Eq. 4.19 represent Joule heating due to electrical current, lateral heat conduction along the plane of a film, and vertical heat conduction through the dielectrics into the substrate, respectively. The contributions by the three terms in Eq. 4.19 are calculated for Cu thin film with different thicknesses of Cu line as shown in Table 4.4. The full substrate thickness of 300 μm is considered and the current density is assumed to be 1 MA/cm^2 in the analysis.

As can be seen in Table 4.4, one can see that the heat conduction through the dielectrics into the substrate increases dramatically as the line thickness decreases. Therefore, it is inaccurate to ignore the last term as it was done in the conventional derivations of Eqs. 4.10–4.12.

Also, in the derivations of Eq. 4.12, the coupling effect between the hydrostatic stress and the temperature is considered and formulated as follows:

$$\sigma_H = \frac{1}{3}\left(\sigma_{11} + \sigma_{22} + \sigma_{33}\right) = -\frac{2E\Delta\alpha_l}{3(1 - v)}(T - T_{SFT}) \tag{4.20}$$

where σ_H is the hydrostatic stress in the interconnect, σ_{11}, σ_{22} and σ_{33} are stresses along the principal axes, T_{SFT} is the stress-free temperature of the interconnect, and $\Delta\alpha_l$ is the difference of the CTE between the interconnect metallization and the surrounding materials. Equation 4.20 is derived based on the Eshelby's inclusion

Table 4.4 The contribution of the three terms in equation

Line thickness (μm)	$J^2\rho$ (pW/μm^3)	$K\nabla^2 T$ (pW/μm^3)	Third term of Eq. 4.19 (pW/μm^3)	Contribution by third term of Eq. 4.19 (%)
3	3.46×10^6	3.26×10^6	0.2×10^6	6.1
1	3.46×10^6	2.84×10^6	0.62×10^6	17.9
0.325	3.46×10^6	1.56×10^6	1.9×10^6	54.9
0.2	3.46×10^6	0.36×10^6	3.1×10^6	89.6

Table 4.5 The hydrostatic stress by Eq. 4.20

Line width (µm)	Equation 4.20 (MPa)	Simulation (MPa)		
		Min	Max	Volume-averaged
1	94.8	61	149	112.1
0.6	94.8	65.2	151	119
0.28	94.8	74.7	159	133
0.16	94.8	78.3	168	140.3

model [25] and the cross section of the metal film was assumed to be an elongated ellipsoid in their analysis.

In reality, the cross section of a typical Cu interconnect is rectangular with cap layer on the top and diffusion barrier layers on the sidewalls and bottom; therefore, Eq. 4.20 is inappropriate in describing the coupling effect between the stress and temperature. In fact, the stress distribution is highly nonuniform due to the material and the structural inhomogeneities in interconnects. Simulations show that the thermo-mechanical stress distribution is strongly dependent on the structure geometry such as the line width, the aspect ratio, and the presence of via. The discrepancy will be more severe for narrow interconnects due to the increased thermo-mechanical hydrostatic stress as shown in Table 4.5. The volume-averaged stress shown in Table 4.5 is calculated using Eq. 4.21

$$\sigma_{volume\_ave} = \frac{\int_V \sigma_H dV}{\int_V dV}. \tag{4.21}$$

The volumetric domain size in computing Eq. 4.21 is chosen to be relatively large such that further increase in the size will only alter the result by <1%. The difference between SFT and test temperature is taken to be 50°C in the analysis.

Therefore, the two assumptions stated above affect the accuracy of the conventional driving force formulation when they are applied to narrow interconnects.

From the equations of the atomic flux, Eqs. 4.7–4.9, the divergence of the respective atomic fluxes can be derived based on the Green's theorem [26] without the above-mentioned assumptions, and they are shown as follows:

$$\text{div}(J_A) = \left(\frac{E_A}{k_B T^2} - \frac{1}{T} + \alpha\frac{\rho_0}{\rho}\right)\frac{N}{k_B T}eZ^*\rho D_0 \exp\left(-\frac{E_A}{k_B T}\right)j \cdot \nabla T \tag{4.22}$$

$$\text{div}(J_{th}) = -\left(\frac{E_A}{k_B T^2} - \frac{2}{T}\right)\frac{NQ^*D_0}{k_B T^2}\exp\left(-\frac{E_A}{k_B T}\right)\nabla T \cdot \nabla T$$
$$-\frac{NQ^*D_0}{k_B T^2}\exp\left(-\frac{E_A}{k_B T}\right)\nabla \cdot (\nabla T) \tag{4.23}$$

$$\text{div}(J_s) = \left(\frac{E_A}{k_B T^2} - \frac{1}{T}\right)\frac{N\Omega D_0}{k_B T}\exp\left(-\frac{E_A}{k_B T}\right)\nabla\sigma_H \cdot \nabla T$$
$$+\frac{N\Omega D_0}{k_B T}\exp\left(-\frac{E_A}{k_B T}\right)\nabla \cdot (\nabla\sigma_H). \tag{4.24}$$

Compared with the conventional AFD formulation in Eqs. 4.10–4.12, the new derivations are without the above-mentioned assumptions. For example, in Eq. 4.23 of the AFD due to TGIDF, vertical heat conduction through the dielectrics into the substrate is considered. For AFD due to SGIDF in Eq. 4.24, the coupling effect between the hydrostatic stress and the temperature is simulated by two coupled-field analysis in ANSYS@, namely the electrical–thermal and the thermal–mechanical analysis, without any assumption on the geometry of the structure. Furthermore, the directions of the respective driving forces are taken into consideration where all the driving forces are represented as vectors and their interactions are represented by the dot product in Eqs. 4.22–4.24.

Tan et al. [15] verified the AFD formulations in Eqs. 4.22–4.24 through the EM modeling in reservoir interconnect structures. The geometry parameters for the finite element models were taken from Refs. [27, 28] as shown in Table 4.6. By virtue of the symmetry of the structure, only half of the model was simulated. The Ta diffusion barrier and SiN cap layer were also included as shown in Fig. 4.9. The material properties were taken from Refs. [2, 15] and listed in Table 4.7. Six finite

Table 4.6 Structural dimension list

Feature	Size (μm)
Line width	0.28
Line thickness	0.35
Via diameter	0.26
Via height	0.68
Barrier layer thickness	0.025
Cap layer thickness	0.05
Reservoir length	0–0.125
Silicon substrate thickness	300

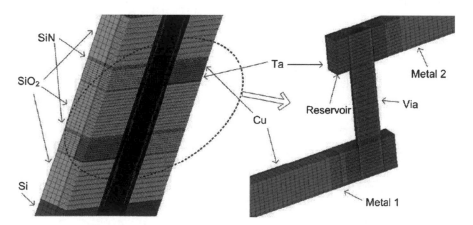

Fig. 4.9 Finite element model for reservoir EM study. Reprinted with permission from Tan et al. [15], copyright © 2007, American Institute of Physics

Table 4.7 Material properties list [2, 15]

Material	Young's modulus (GPa)	Poisson ratio	Thermal conductivity (W/mK)	Coefficient of thermal expansion (/°K)
Cu	129.8	0.339	379	16.5×10^{-6}
Ta	186.2	0.35	53.65	6.48×10^{-6}
SiN	265	0.27	0.8	1.5×10^{-6}
SiO$_2$	71.4	0.16	1.75	0.68×10^{-6}
Si	130	0.28	61.9	4.4×10^{-6}

element models with different reservoir lengths spanning from 0 to 0.125 μm were constructed. These EM models are subjected to a current density of 1.2 MA/cm^2 and a test temperature of 300°C. The SFT is assumed to be 350°C [13].

4.4.2 Finite Element Model of EM Using Green Theorem-Based AFD Formula

4.4.2.1 Void Nucleation (Static Simulation)

The finite element models developed consists of two steady-state analyses, namely a direct coupled analysis with the current and the temperature fields, and an analysis with indirect coupling between the temperature and the stress fields. The flow chart for the static EM simulation is shown in Fig. 4.10.

In electrical–thermal analysis, the substrate bottom surface was kept at constant EM test temperature with constant current density applied. The nodal temperatures were then retrieved from this stage for the next thermal–mechanical analysis. In the thermal–mechanical analysis, the following boundary conditions were applied:

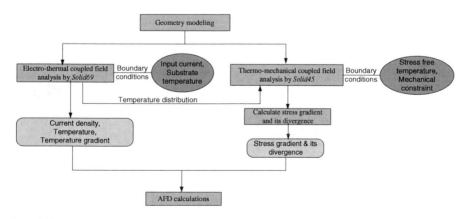

Fig. 4.10 Flow chart for the static EM simulation

(1) the substrate bottom was fixed with zero displacement and (2) the vertical symmetric plane was constrained to remain vertical arising from the mirror symmetrical nature of the structure under consideration.

User subroutines were developed to calculate the divergence of temperature gradient and thermo-mechanical stress gradient in ANSYS$^{@}$. The AFD contributions due to EWF, TGIDF, and SGIDF were calculated using Eqs. 4.22–4.24 for the Green theorem-based method, and Eqs. 4.7–4.9 for the conventional driving force method. The location of maximum positive AFD was taken as the void nucleation site [29].

Figures 4.11 and 4.12 show the AFD distributions of the M2 structure using the Green theorem-based method and the conventional formulations, respectively. For easy visualization, all the other materials are removed in the figures and only the bare Cu is shown. One can see that the predicted void nucleation sites are different

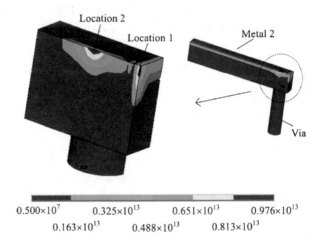

Fig. 4.11 AFD distributions using the proposed FEM and the maximum AFD sites are labeled as "Location 1" and "Location 2" at Cu/SiN interface. Reprinted with permission from Tan et al. [15], copyright © 2007, American Institute of Physics

$$0.500\times10^{7} \qquad 0.325\times10^{13} \qquad 0.651\times10^{13} \qquad 0.976\times10^{13}$$
$$0.163\times10^{13} \qquad 0.488\times10^{13} \qquad 0.813\times10^{13}$$

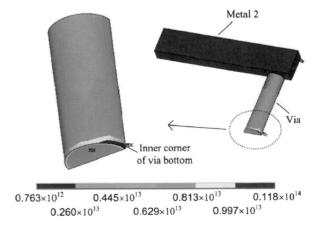

Fig. 4.12 AFD distributions using the conventional FEM. The maximum AFD site is found to be at inner corner of via bottom. Reprinted with permission from Tan et al. [15], copyright © 2007, American Institute of Physics

$$0.763\times10^{12} \qquad 0.445\times10^{13} \qquad 0.813\times10^{13} \qquad 0.118\times10^{14}$$
$$0.260\times10^{13} \qquad 0.629\times10^{13} \qquad 0.997\times10^{13}$$

Table 4.8 AFD contributions of different driving forces

Driving force	Atomic flux (atoms/μm^2 s)	Atomic flux divergence (atoms/μm^3 s)
EWF	6.54×10^{11}	1.34×10^8
TGIDF	4.45×10^8	1.83×10^9
SGIDF	7.36×10^{13}	5.69×10^{12}

between these two methods. Conventional method predicts the void nucleation site at the inner corner of the via bottom as indicated in Fig. 4.12. However, via bottom failure is proven to be the early failure due to process-related defects [30] and it can be eliminated by via process optimization [31]. Hence, the prediction by the conventional formulation is incorrect in the defect-free model.

On the other hand, two maximum AFD sites labeled as "Location 1" and "Location 2" is predicted by the Green theorem-based method as shown in Fig. 4.11, and both "Location 1" and "Location 2" are at the Cu/SiN interface. "Location 1" is located at the corner of the M2 line corresponding to the site of the maximum thermo-mechanical stress [32]. "Location 2" is located directly above the via and Vairagar et al. proved that it is indeed the void nucleation site using qualitative Monte Carlo simulation [33]. In fact, the two potential void nucleation sites are well reported in the literatures [32, 34]. Table 4.8 summarizes the contributions by the various driving forces for the atomic flux (AF) and AFD, and SGIDF is identified as the dominant driving force which also agrees with the studies by Shen et al. [35] and Tan et al. [19].

4.4.2.2 Void Growth (Dynamic Simulation)

After void nucleation, the voids begin to grow due to the continuous removal of materials from the void nucleation sites. The AFD value represents the rate of mass transport. Figure 4.13 shows the flowchart of the dynamic simulation. Twenty elements with the highest total AFD values are selected and physically deleted in each loop to simulate the void growth. The geometry of the FE model is then modified and this process repeats itself until the interconnect resistance increases by 10%. Figure 4.14 shows the FE model after an iteration of 20 loops. It can be clearly seen that the voids grow simultaneously at "Location 1" and "Location 2" at the beginning of the void-growth process, in agreement with the experimental results reported in Ref. [36].

4.4.2.3 Critical Reservoir Length Estimation

It is well-known that the introduction of reservoir as shown in Fig. 4.15 can enhance the EM performance. However, it is also observed that the lifetime improvement ceases when the length of the reservoir is above a critical value. Gan et al. [27] introduced a concept of "effective reservoir volume" where only

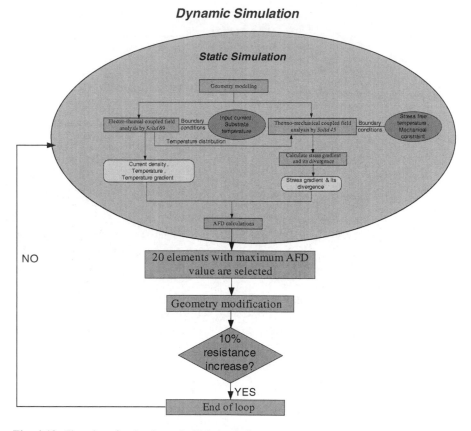

Fig. 4.13 Flowchart for the dynamic EM simulation

Fig. 4.14 Void growth at "Location 1" and "Location 2" simultaneously after 20 iteration loops. Reprinted with permission from Tan et al. [15], copyright © 2007, American Institute of Physics

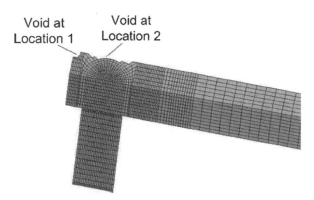

part of the extension volume is served as the reservoir for void accumulation. Recently, Fu et al. [37] proved that the effective reservoir volume concept cannot explain the reservoir enhancement effect as observed in their experimental.

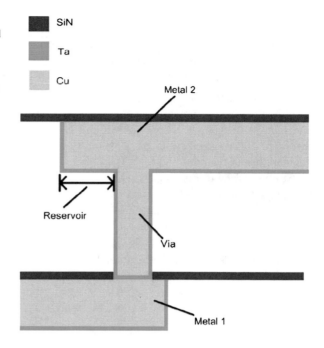

Fig. 4.15 Schematic diagram of the cross sectional view of the reservoir region. Reprinted with permission from Tan et al. [15], copyright © 2007, American Institute of Physics

In order to explore the mechanism of the reservoir effect using FEM, we examine the lifetime from the AFD calculations. The EM lifetime is inversely proportional to the volume-averaged AFD as can be derived below. In other words, AFD can be employed to predict the EM lifetime qualitatively.

Consider a control volume equal to the size of the critical volume of the void that causes 10% resistance increase in interconnects, i.e. the void volume at EM failure. If $\overline{AFD_{\text{total,V}}}$ is the volume-averaged of the total AFD within the control volume, then by virtue of the conservation of mass, we have

$$\overline{AFD_{\text{total,V}}} + \frac{\delta N}{\delta t} - \frac{\delta C}{\delta t} = 0 \tag{4.25}$$

where C is the average vacancy concentration and N is average atomic concentration within the control volume, respectively. Since the vacancy contribution is at least five orders of magnitude smaller than the atomic contributions for the void formation, it can be ignored [38].

Integrating Eq. 4.25 gives

$$\int_{N_o}^{0} dN = -\int_{0}^{TTF} \overline{AFD_{\text{total,V}}}\, dt \tag{4.26}$$

$$N_o = \overline{AFD_{\text{total,V}}} \times TTF$$

where $\overline{\overline{AFD}}_{\text{total,V}}$ is the $\overline{AFD}_{\text{total,V}}$ averaged over the duration of TTF, and it reflects the average material depletion rate in the void-growing process. TTF is the time to failure, i.e. the EM lifetime. Since N_o is the initial atomic concentration of the metal and is a constant for a given metal, we have the EM lifetime inversely proportional to the $\overline{\overline{AFD}}_{\text{total,V}}$ as shown in Eq. 4.26.

When a void is formed inside a metal line, partial relaxation of the thermo-mechanical stress in the vicinity of the void is evidenced [35]. Further relaxation of stress can be realized by either plastic deformation or diffusion. Since EM test temperature is relatively high at 300°C, the thermo-mechanical stress is assumed to be released by diffusion [25], which leads to an increased SGIDF subsequently. As a result, the total AFD increases.

Figure 4.16 shows the variation of $\overline{AFD}_{\text{total,V}}$ within the control volume after each loop of iteration, depicting the progressive AFD growth as voids are growing.

The fluctuation of the AFD in Fig. 4.16 is the consequence of the finite size of the elements used for the element deletion in the void growth simulation. As was derived earlier, one will take the mean of the volume-averaged total AFD, $\overline{\overline{AFD}}_{\text{total,V}}$ for analysis as it is related to the EM lifetime, and its value is shown in Fig. 4.16.

Figure 4.17 shows the $\overline{\overline{AFD}}_{\text{total,V}}$ for different reservoir lengths. One can clearly see that the $\overline{\overline{AFD}}_{\text{total,V}}$ decreases gradually with the reservoir length until a critical point of around 0.08 µm, after which the $\overline{\overline{AFD}}_{\text{total,V}}$ remains constant. In other words, the void growth rate decreases as the reservoir length increases and correspondingly the EM lifetime increases till around 0.08 µm. Further increase in the reservoir length does not decrease void growth rate $\overline{\overline{AFD}}_{\text{total,V}}$, making the lifetime enhancement effect of the reservoir length ceased. The result is in close agreement with Gan et al. [27] who found that the critical length is around 0.06 µm using the same test structure and test conditions studied here. On the other hand, the conventional formulation predicts the voids at via bottom, and hence no reservoir effect can be predicted by the conventional method.

Fig. 4.16 Volume-averaged AFD, $\overline{AFD}_{\text{total,V}}$ within the control volume during void-growing process. Reprinted with permission from Tan et al. [15], copyright © 2007, American Institute of Physics

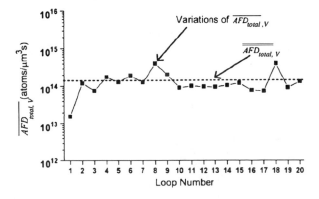

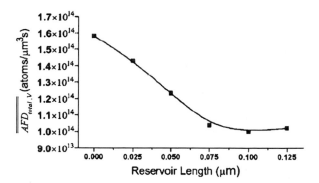

Fig. 4.17 Variation of the mean of the volume-averaged AFD, $\overline{AFD_{total,V}}$, within the control volume with the reservoir length. $\overline{AFD_{total,V}}$ decreases gradually with the reservoir length until a critical length of around 0.08 μm. Reprinted with permission from Tan et al. [15], copyright © 2007, American Institute of Physics

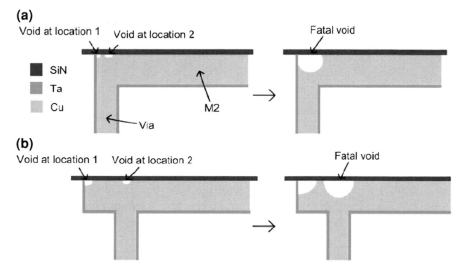

Fig. 4.18 Void-growing process in the reservoir region. **a** For Metal-2 structure with no reservoir extension, the initial two voids collapse into a bigger void and a critical void forms above via finally. **b** For Metal-2 structure with large reservoir extension, the critical void is solely developed from the initial void at "Location 2". Reprinted with permission from Tan et al. [15], copyright © 2007, American Institute of Physics

4.4.2.4 Mechanisms of the Reservoir Length Effect

From the above modeling results, we found that the dominant driving force for the void formation is SGIDF in the reservoir region. High thermo-mechanical stress gradient is responsible for the void nucleation and the subsequent void growth is through stress relaxation in the form of diffusion. Current crowding effect is ignored in the analysis because the reservoir is a low current density region. Besides, it is also shown numerically that the effect of current density gradient driving force is negligible [27].

Figure 4.18 explains the mechanism of reservoir effect schematically. In the case of the test structure without any reservoir extension as shown in Fig. 4.18a,

"Location 1" and "Location 2" grow simultaneously in the beginning. After some time, the two voids collapse into a bigger void and form a critical void above via finally.

However, in case of the test structure with a reservoir extension, the contribution by the void at "Location 1" to the final critical void decreases with increasing reservoir length due to the increased distance between "Location 1" and the via, and as a result, the $\overline{AFD}_{\text{total,V}}$ decreases. When the reservoir length increases beyond a critical value as shown in Fig. 4.18b, the void at "Location 1" will not be able to collapse with the critical void above via before the line fails. In other words, the critical void grows solely from "Location 2", hence further increase in the reservoir length will not enhance the EM lifetime.

In short, the underlying physics of the reservoir effect is the decreasing probability of the void in the reservoir extension to be part of the critical void for EM as the reservoir length increases.

4.4.3 Summary

We have discussed the assumptions employed in the conventional AFD formula, and the inaccuracies of the formula for narrow interconnects. A refined driving force FEM is proposed based on the Green theorem, and the validity of the method is demonstrated by applying the method to investigate the void nucleation sites, the void growth process, and the reservoir effect on EM in copper DD interconnects. From the investigation, it is found that thermo-mechanical stress-induced migration is the dominant driving force for void nucleation and growth in the M2 reservoir region.

One of the drawbacks of the FEM simulation methodology is its incapability in simulating the activity of vacancies/atoms during EM from a thermodynamics aspect, i.e. the diffusion paths of the vacancies/atoms movement cannot be included in the study. However, the redistribution of vacancies/atoms through various diffusion paths during the void nucleation at the early stage of EM failure is essential for the narrow interconnect system. This is because of the small dimension and the high current density in the narrow interconnect system, and the process of EM failure after the void is nucleated can be very abrupt [39]. Under such circumstances, the method to retard the void nucleation at the early stage of EM failure will be a primary concern as this period of time could be almost the entire EM lifetime for narrow interconnections.

The redistribution of vacancies/atoms will not disturb the physical environment of the entire EM testing structure significantly, and the process is nearly impossible to be simulated using the AFD simulator described above since the calculation of AFD is based on the finite element modeling on the physical environment of the interconnect. In the following section, a promising simulation methodology for the void nucleation will be introduced and its incorporation into current EM AFD simulator is also described.

4.5 Monte Carlo Method for Electromigration Simulation

Monte Carlo method is a class of nondeterministic computational algorithms for simulating the behavior of various physical and mathematical systems. There are only few literatures on the simulation of EM using Monte Carlo method.

The Monte Carlo method is different from other simulation methods by being stochastic as opposed to deterministic algorithms introduced in the above sections. In this section, we first present a brief review on a few applications of Monte Carlo method in the simulation of the physical process of EM, and then we will discuss a methodology to combine the Monte Carlo method and FEM to form a holistic EM simulation for void nucleation and void growth.

4.5.1 The Application of Monte Carlo Method in EM Study

4.5.1.1 Monte Carlo Simulation on the Movement of Atoms During EM

Smy et al. reported their work on Monte Carlo computer simulation of EM in 1993 [40]. In the model, the motion of atoms was simulated according to two processes during EM, namely a drift of activated atoms in response to the EWF and the annealing process in metal line due to the atoms movement to minimize the curvature of the grain boundaries. The grain structure was modeled by disks, each comprising some 2,000 atoms. This was an attempt to describe events at the atomic level limited by the computer memory. However, the physical laws used were not those obeyed by an individual atom, but rather were those appropriate to describe the average behavior of the atoms. The simulation of the EWF migration consisted of two steps, the selection of disks at random and the drifting of disks under EWF. The number of disks selected was proportional to the number of activated atoms given below.

$$N_1 \approx e^{-E_{gb}/2kT} N_{gb} \tag{4.27}$$

where N_{gb} is the number of grain boundary disks, and E_{gb} the activation energy of activated atom formation and its motion. They were drifted under EWF according to the following equation [40]

$$v = \frac{Z^* e}{k_B T} \exp\left(-E_{gb}/2k_B T\right) J \cos\phi \sin\left(\frac{\theta}{2}\right), \tag{4.28}$$

where θ is the misorientations of grains and ϕ is the angle between the grain boundary and the prevailing current density as indicated in Fig. 4.19.

Annealing is a motion of disks to minimize surface curvature, and is carried out at a fraction of the rate at which the disks are migrated. The disks were also selected at random along grain boundary and were moved from the sites of positive

Fig. 4.19 The Monte Carlo simulation of electromigration and annealing event with the consideration of grain orientation and current flow direction. Reprinted with permission from Smy et al. [41], Copyright © 1993, American Institute of Physics

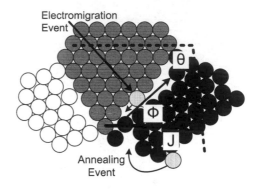

Fig. 4.20 Atoms and sites for two grains of different orientations. Reprinted from Computational Material Science, Bruschi et al. [44], copyright © 2000, with the permission from Elsevier

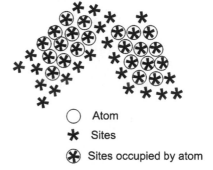

curvature to the sites of negative curvature for reducing the surface energy. The different patterns of the disk represent different grains. The motion of disks during the EM and the annealing events are illustrated in Fig. 4.19.

These disk manipulation routines were collectively referred to as SIMBAD code, reported in their previous reports, which have been applied to the studies in areas as diverse as the microstructure of deposited metals [41], dielectrics [42], and the formation of hailstones [43].

Another atomic Monte Carlo simulation of EM in polycrystalline thin films was reported by Bruschi [44]. In their study, they proposed an atomic Monte Carlo EM simulations based on a model of atomic migration that includes the effect of EWF on the activation barrier. The samples were represented by a 3D array of cubic cells. Each cell was a data structure which includes the following objects: (1) a list of sites, (2) a list of atoms, and (3) an EWF. The EWF was calculated by simulating the injection of a constant current and solving an equivalent resistor network extracted from the sample. The atomic Monte Carlo simulator calculated the interatomic interactions and sought the possible destination for atom hopping which can be either a substrate site or an induced site (Fig. 4.20).

Substrate sites simulated the vacancies in a grain, and induced sites simulated the vacancies along the grain boundary. Incorporation of atoms into the induced-sites result in grain growth. The sample then evolved through a series of transitions with each transition consisting of an atom jump from an initial position to an

Fig. 4.21 Possible cases of void location at the dielectric/metal bamboo film interface (the schematic represents the *top view* of metal line). Reprinted with permission from Zaporozhets et al. [48], copyright © 2005, American Institute of Physics

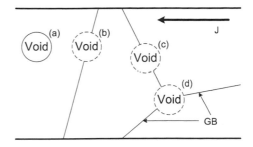

empty site (substrate site or induced site) provided that the site lies within a maximum jump length from the initial position, and the model can simulate the void growth, the hillock formation and the grain growth on the sample.

The atomistic Monte Carlo simulation is a promising approach that can be applied to various problems related to diffusion and nucleation of atoms on a surface [45, 46].

4.5.1.2 Monte Carlo Simulation of Void Movement During EM

With the interconnect technology shifting to Cu, there are new experimental observations of EM reported in the literature, such as void migration at the interface between thin metallic film and dielectric [47]. Zaporozhets et al. [48] proposed an atomistic Monte Carlo methodology to simulate the void migration in a Cu interconnect as observed in their experiments.

The simulator modeled the void shape evolution and atomic displacements using stochastic Monte Carlo method, with probability of displacement depends exponentially on the change of the energy as a result of the displacement. The EWF energy and the pair interaction energy were taken into consideration. The metallic film material was modeled at an atomic level, and the electric field was solved by using finite difference method with a time-dependent boundary condition. The interactions of bulk, grain boundary, dielectric, and vacancy with the moving atoms were described by a matrix of interaction coefficients for different pair interaction energies.

Based on their model, they simulated the behavior of the void under several conditions, namely (1) surface void migration along the interface of metal/dielectric in the absence of grain boundary; (2) void migration along grain boundaries of different orientations with respect to current; and (3) behavior of the voids at triple junctions of grain boundaries (Fig. 4.21).

4.5.2 A Holistic EM Modeling Using Monte Carlo Method and Finite Element Method

While the surrounding materials can induce different driving forces for the diffusion of metal atoms in interconnections and this can be accounted for using FEM, the presence of the various diffusion paths in interconnections also determines the metal atom diffusion rate. Hence, both the driving forces and diffusion paths should be considered in the modeling of EM.

Therefore, a more realistic EM model should include both the diffusion and the driving force approaches with Monte Carlo simulation for the initial void nucleation process. This section will introduce a new methodology for a holistic EM simulation in interconnections. With such a model, the entire EM failure process beginning from void nucleation to its growth can be modeled and the key material properties that influence the reliability of ULSI interconnections can be identified.

4.5.3 Description of Simulation Methodology

4.5.3.1 Test Structure Description

To demonstrate the simulation methodology, Al interconnect test structure is chosen as an illustration. The same principle and methodology can be applied to other metallic interconnect.

The test structure under study is an Al metal line sandwiched between two TiN layers, and the entire metallization stack is embedded in SiO_2 dielectric as shown in Fig. 4.22. The stress-free temperature in the model is set at 410°C which is the final annealing temperature of the wafers [49].

As a common practice, SiO_2 and TiN are taken as isotropic linear elastic solids, and Al is characterized as an isotropic elastic-perfectly plastic solid [50]. These assumptions are justified by Greenabaum et al. [50] in their experiments and calculations. The properties of the materials involved are summarized in Tables 4.9 and 4.10.

Fig. 4.22 Finite element model of Al test structure. Reprinted with permission from Li et al. [62], copyright © 2007, American Institute of Physics

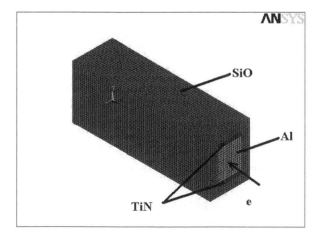

Table 4.9 Material properties used in FEM [2, 40]

Physical parameter	Symbol	Value (Al)
Atom concentration	N	$6.03 \times 10^{10}/\mu m^3$
Effective valence charge	Z^*	10
Pre-exponential factor	D_0	$2.3 \times 10^8 \mu m^2/s$
Resistivity	ρ_0	2.65×10^{-14} TΩ μm
Temperature coefficient of resistivity	α	0.0043
Heat of transfer	Q^*	1.22×10^{-8} pJ
Poisson ratio	v	0.34
Young's modulus	E	7×10^4 MPa
Coefficient of thermal expansion	α_l	23.1×10^{-6}/K
Activation energy for lattice diffusivity	$E_{a,\text{lattic}}$	1.41 eV
Activation energy for grain boundary diffusivity	$E_{a,\text{GB}}$	0.48 eV
Activation energy for surface diffusivity	$E_{a,\text{surf}}$	0.84 eV

Table 4.10 Mechanical properties of Al and its surrounding materials

Mechanical properties	Al[a]	TiN[b]	SiO$_2^c$
Coefficient of thermal expansion (K^{-1})	23.1×10^{-6}	9.4×10^{-6}	0.68×10^{-6}
Young's modulus (MPa)	7×10^4	2.7×10^5	7.1×10^4
Yield strength (MPa)	200 (20°C)	–	–
	67.4 (400°C)		
Poisson ratio	0.35	0.25	0.16
Thermal conductivity (pW/μm K)	2.32×10^8	2.61×10^7	1.75×10^6
Resistivity (TΩ μm)	2.65×10^{-14}	1.24×10^{-12}	1×10^{26}

[a] References [2, 51]
[b] References [10, 52, 53]
[c] References [10, 49]

The dimension of the metal line in the model is 0.475 μm × 0.3 μm × 1.875 μm. Since the analysis involves complex electrical–thermal–mechanical interactions, two coupled-field analyses are employed, namely the electrical–thermal coupled-field analysis and the thermal–mechanical coupled-field analysis. Element *Solid 69* with tetrahedral shape is used in the electrical–thermal analysis, and *Solid 45* is used for the thermal–mechanical analysis. All the elements have an identical size of 0.025 μm × 0.025 μm × 0.025 μm. Different grains in the thin film are created in a random manner as shown in Fig. 4.23. The elements along grain boundary possess different atomic diffusivity and adhesion energy from those elements inside the grain. The details of the energy calculation will be presented later.

The stress current density is 3 MA/cm^2 and the stress temperature is 200°C. The temperature distribution obtained from the electrical–thermal analysis, including the Joule heating effect, is applied as loads to *Solid 45* model for the thermal–mechanical analysis.

Fig. 4.23 Different grains in the metallization. Surrounding materials are not shown for the sake of clarity. Reprinted with permission from Li et al. [62], copyright © 2007, American Institute of Physics

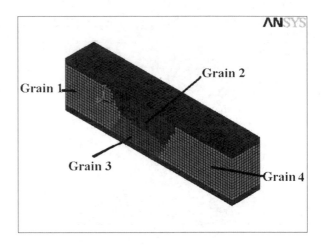

4.5.3.2 Monte Carlo Simulation of Void Nucleation

Void nucleation is a process of motion of vacancies and atoms in order to minimize the system energy. To model the process of void nucleation qualitatively, Monte Carlo method is employed. Three aspects are considered in the simulation as follows, namely energy calculation, element movement, and Monte Carlo subroutine.

4.5.3.2.1 Energy Calculation

In the presence of the driving forces during EM, the total energy of an element consists of the EWF energy, the strain energy, the thermal energy, and the pairing energy between the element and its nearest neighboring elements.

The pairing energy represents the initial cohesion between one element and its surrounding elements without any external excitations such as elevated temperature and stress. In the absence of any external excitations, an element is at its lowest energy state. The pairing energy is the amount of the energy required to bring the element from its lowest energy state and free it out of its surrounding elements. It is estimated by using the relation $E_{\text{pair}} = -zE_f$, [48, 54] where z is the number of the nearest neighbor elements, E_f is the adhesion energy between two elements. The minus sign in the expression of the pairing energy indicates that if E_f or z is higher, the pairing energy state of an element will be more negative, and higher excitation energy is required to free the element from its surrounding elements. In other words, the element will be more stable in its position. Within an Al grain, E_f is estimated from the sublimation energy of Al and its value is approximately 2.487×10^{-4} pJ [48].

The adhesion energy between TiN and Al is $E_{f\_\text{TiN}} \approx 3 \times E_f$ [55]. Due to the formation of an ultrathin alumina between SiO_2 and Al [55], the adhesive energy of Al and SiO_2, E_{f_SiO}, is the adhesive energy of Al/Al_2O_3 and it is estimated as $E_{f\_SiO} \approx E_f$ [56]. For the elements in different Al grains, the adhesion energy is estimated as $E_{f\_GB} \approx 0.4 \times E_f$ [48].

By using different pairing energies for the elements along the grain boundaries and inside the grains, we simulate the effect of inhomogeneity due to the interconnect microstructure and texture on EM.

The strain energy and the thermal energy are two external excitations on an element. They both raise the energy state of the element from its initial state which is defined by the pairing energy described previously. Explicitly, the probability of movement of the element is higher due to these two external excitations, since the minimum energy required for the movement of the element is lowered. In other words, the pairing energy, the strain energy and the thermal energy simulate the stability of an element, and the summation of these three energies represents the cohesion of the element with its surrounding elements.

The strain energy is given as [57]

$$E_{str} = E_e^{el} + E_e^{pl} = \frac{1}{2}\{\sigma\}^T\{\varepsilon^{el}\}\text{vol}_e + \{\sigma\}^T\{\Delta\varepsilon^{pl}\}\text{vol}_e. \tag{4.29}$$

Here E_e^{el} and E_e^{pl} are the elastic and plastic strain energy, respectively. $\{\sigma\}$ is the stress vector, $\{\varepsilon^{el}\}$ is the elastic strain vector, $\{\Delta\varepsilon^{pl}\}$ is the plastic strain increment, vol_e is the volume of the element.

The thermal energy is given as

$$E_{th} = N \times \frac{3}{2}k_B T, \tag{4.30}$$

where N is the number of the atoms in one element, T the temperature of the element in Kelvin, and k_B the Boltzmann's constant.

The third external excitation is the electron current flow. The EWF energy is equal to the work done by the EWF during an element displacement, and it is given as $E_{ew} = Z^*eU$ [48], where Z^*e is the effective charge of Al atom, U is electrical potential.

4.5.3.2.2 Element Movement

With the temperature and current density used in the present work, the strain energy, and the thermal energy are not high enough to balance or exceed the pairing energy of an element (a case of de-cohesion), but the energy required to free an element is reduced under these excitations, and hence the probability of an element movement increases. Together with the excitation from the EWF, the probability of the movement of an element becomes higher along a particular direction. Based on the statistical mechanics, the probability of an element displacement is exponentially dependent on the change of the total energy between the target position and the original position of the element. The probability is given as [40, 48, 54]

$$\text{Probability} = \exp\left(-\frac{\Delta E_{\text{ew}} + \Delta E_{\text{str}} + \Delta E_{\text{th}} + \Delta E_{\text{pair}}}{N \times k_B T}\right). \tag{4.31}$$

4.5.3.2.3 Monte Carlo Subroutine

Monte Carlo method is used to simulate the movement of an element from one lattice location to another. To begin, groups of vacancy sites, which are empty without elements in the finite element model, are randomly generated along the grain boundaries. Since the grain boundary is the main diffusion path of EM in Al polycrystalline thin film, the movement of the vacancies that reside along these grain boundaries is the focus. The reported average vacancy concentration in Al is found to be 0.01–0.1% [58, 59] and the vacancy concentration of 0.1% is chosen in this study because the concentration is expected to be higher along the grain boundaries at the elevated test temperature of 200°C.

A Monte Carlo subroutine randomly selects one vacancy site, and the total energy change for each of its nearest neighbor elements to move and fill this vacancy site is computed individually. The movement of an element is accepted if its probability as computed using Eq. 4.31 is the highest among other surrounding elements and above 50% [54]. The particular element and vacancy site will then swap their original positions, and a new Monte Carlo loop begins with another random selection of a vacancy site. After a number of Monte Carlo loops, the movement of vacancies in the structure can be observed, and the void nucleation as a result of the vacancies clustering can be identified.

4.5.3.3 Dynamic Simulation of Void Evolution

After the void nucleation, the values of AFD at the nucleation sites due to the various driving forces are calculated, and the voids begin to grow due to the continuous removal of material from the nucleation sites with the rate of the mass transport given by the value of the total AFD as described in Sect. 4.4. Different diffusion paths are taken into account by considering the different diffusivities of metal atoms along the grain boundaries, the interfaces, and inside the grains. The activation energies of the diffusivities for the different diffusion paths are included in the AFD computation as listed in Table 4.9.

4.5.4 Result and Discussion of the Holistic Modeling

Figures 4.24, 4.25, 4.26, 4.27, 4.28 and 4.29 show the initial and final profiles of the vacancies in the metal line in four different viewing angles. As shown in the figures, the vacancies move through the grain boundaries from the anode to the cathode, and ultimately become immobilized at the intersection of the barrier layer

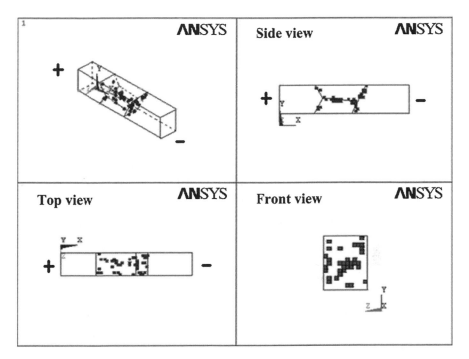

Fig. 4.24 Randomly generated vacancies along the grain boundaries. Reprinted with permission from Li et al. [62], copyright © 2007, American Institute of Physics

and the grain boundary. The immobilization occurs because the intersection has higher strain energy due to thermal mismatch of the different materials at the intersection. Vacancies tend to move and nucleate there so that the strain energy can be reduced.

Vacancies also nucleate at other locations inside the structure as shown in Fig. 4.27, such as the grain boundaries or sidewalls as these are also weak points in the metal line. However, the size of the nucleated voids is relatively small at these weak points. This is in agreement with the experimental observations reported by Smith et al. [56]. They found that some of the nucleated voids cannot grow and form fatal voids across the metal line eventually. As shown in Fig. 4.28, the nucleated voids located at the barrier layer–grain boundary intersection grow at a faster rate due to higher AFD when compared to the voids at other weak points in the interconnect. These nucleated voids grow, merge, and finally form a fatal void which covers almost the entire metal cross-section and causes the interconnect resistance to increase sharply.

Figures 4.27 and 4.28 show the void growth process, and the growth process is similar to the observation reported by Prybyla et al. [60]. In their experiment, they used an in-situ transmission electron microscopy (TEM) to record the process of the mini-void migration and they observed the trapping of the voids at the barrier layer–grain boundary intersection. These trapped voids then began to grow along

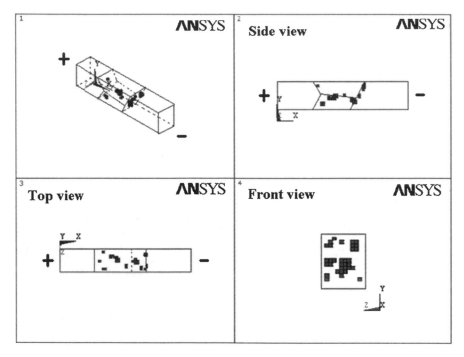

Fig. 4.25 Vacancy movement along the grain boundaries. Reprinted with permission from Li et al. [62], copyright © 2007, American Institute of Physics

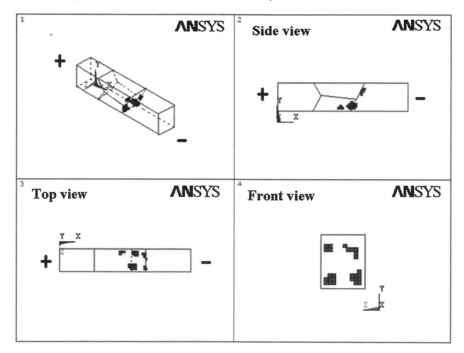

Fig. 4.26 Nucleation of the voids inside the metal line. Reprinted with permission from Li et al. [62], copyright © 2007, American Institute of Physics

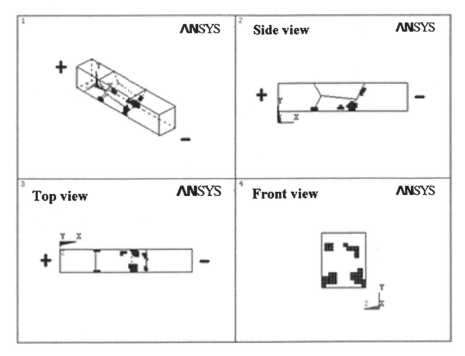

Fig. 4.27 Void growth in the metal line. Reprinted with permission from Li et al. [62], copyright © 2007, American Institute of Physics

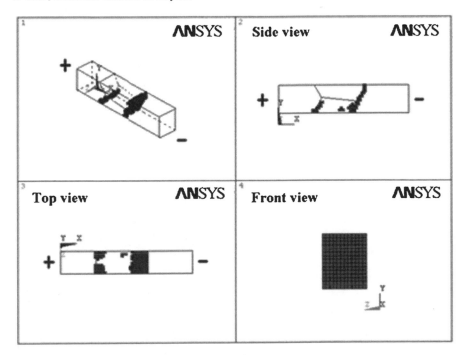

Fig. 4.28 Formation of the fatal void across the metal line. Reprinted with permission from Li et al. [62], copyright © 2007, American Institute of Physics

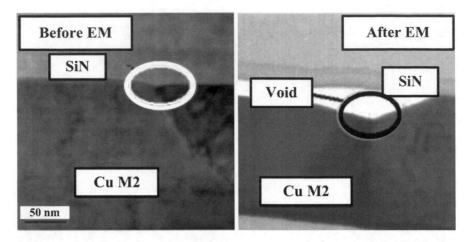

Fig. 4.29 Experimental observation of the void at the capping layer–grain boundary intersection in Cu interconnections after EM (courtesy from Prof Subodh Mhaisalkar). Reprinted with permission from Li et al. [62], copyright © 2007, American Institute of Physics

the grain boundary. The Monte Carlo simulation agrees well with their experimental observation.

The capping layer–grain boundary intersection is also found to be the weak point in Cu interconnections during EM test [61]. In recent Cu interconnect technology, special treatment of the interface between the capping layer and Cu thin film is employed to retard the Cu atoms migration along the fast interface diffusion path effectively. However, due to the thermal mismatch of the two materials at the capping layer-grain boundary intersection, the intersection poses a potential reliability hazard. Figure 4.29 shows a Cu EM sample which is fabricated with SiH_4 surface treatment at the Cu/capping layer interface. The SiH_4 treatment increases the bonding strength of the top interface and consequently reduces the mass transport along the Cu/capping layer interface. Upon failure analysis on the EM-failed sample, it is found that the EM-induced voids occur at the capping layer–grain boundary intersection of Cu metal line. Such a failure phenomena could be due to the thermal mismatch of different materials at the intersection, similar to the simulation result in an Al metal line.

In short, the high defect density at grain boundaries and the thermal mismatch of different materials at the intersections of barrier layers and grain boundaries make the intersections to be vulnerable weak sites in both Al and Cu interconnections.

4.5.5 Summary

A methodology that combines Monte Carlo algorithm and FEM is presented to study the dynamic physical process of EM. The inhomogeneity of the interconnect microstructure and the resulting different atomic diffusivities along the various diffusion paths are considered in the model. The presence of the various driving forces of EM is also included in the analysis. From the modeling, the capping layer–grain boundary intersection is found to be the weakest point for the void formation in the metallization due to the high stress resulting from the thermal mismatch of different materials at the intersection. The model demonstrates the vacancy trapping, void nucleation, and void growth during EM, and the results are found to agree well with the experimental observations reported in the literatures.

4.6 Conclusion

In this chapter, we have discussed in detail the methodology for a 3D simulation of EM based on FEM. The EM kinetic due to various driving forces, such as EWF, TGIDF, and SGIDF, is quantitatively modeled through the calculation of the corresponding AFD. Instead of solving the EM equation in a multi-dimension model, the necessary physical parameters are firstly calculated using FEM. The AFD due to different EM-driving forces is then calculated locally. With the knowledge of AFD at every point in the model, the formation and evolution of the void can be simulated. Using this simulation methodology, the EM kinetic can be evaluated not only in the voiding region, but also within the entire EM testing structure.

The finite element 3D model is able to simulate the effect of the surrounding material, the geometry of the metal line and the metal stack structure in the interconnect system. In comparison with the 1D EM simulation techniques, the EM simulator using AFD through FEM is capable of studying the EM kinetics locally around the region of voiding, and addressing the EM performance and potential reliability weakness of a metal interconnect from a microscopic point of view.

On the other hand, the use of Monte Carlo methods to model physical problems allow us to examine more complex systems. The EM simulation using Monte Carlo method is based on a single atom or a single vacancy. With the basic theory of EM physics, Monte Carlo method randomly generates numbers of scenarios for each atom/vacancy to simulate its behavior. The EM process is modeled by tracking the behavior of all atoms, vacancies, or their clusters. In fact, the accuracy of a Monte Carlo simulation is proportional to the square root of the number of scenarios used. Unfortunately, this also means that it is computing-intensive and should be avoided if simpler solutions are possible. While solving equations which describe the interactions between two atoms is fairly simple, solving the same equations for hundreds and thousands of atoms is nearly impossible for a standard computer system. For this reason, Monte Carlo method does not become the main stream of EM simulation.

However, as demonstrated by Li et al. [62] and as discussed in Sect. 4.5, the Monte Carlo methodology can be modified to simulate the behavior of vacancy clusters without compromising the accuracy of the simulation. The Monte Carlo simulation can actually be incorporated with FEM to simulate the void nucleation from the thermodynamics perspective during EM process as well as the effect of microstructures of the metal line. In this way, both the EM kinetics (driving forces) and its thermodynamics aspect (diffusion paths) which are essential in the study of the underlying physics of EM process can be simulated. In the application of nano-interconnect system in integrated circuit, the understanding and the simulation of void nucleation is increasingly important, and such void nucleation is best studied using Monte Carlo simulation from the thermodynamic aspects. The attempt of combining the simulation of EM kinetics and its thermodynamic aspects will produce a more realistic EM model and help us to understand the physical phenomenon of EM better.

References

1. Kondo S, Hinode K (1995) High-resolution temperature measurement of void dynamics induced by electromigration in aluminum metallization. Appl Phys Lett 67:1606
2. Tan CM, Zhang G, Gan ZH (2004) Dynamic study of the physical process in the intrinsic line electromigration of deep-submicron copper and aluminum interconnects. IEEE Trans Dev Mater Reliab 4:450
3. Sasagawa K, Nakamura N, Saka M, Abe H (1998) A new approach to calculate atomic flux divergence by electromigration. Trans ASME J Electron Pack 120:360
4. Sasagawa K, Naito K, Saka M, Abe H (1999) A method to predict electromigration failure of metal lines. J Appl Phys 86:6043
5. Sasagawa K, Nakamura N, Saka M, Abe H (2002) Governing parameter for electromigration damage in the polycrystalline line covered with a passivation layer. J Appl Phys 91:1882
6. Huntington HB, Grone AR (1961) Current-induced marker motion in gold wires. J Phys Chem Solids 20:76–87
7. Lloyd JR, Smith PM, Prokop GS (1982) The role of metal and passivation defects in electromigration-induced damage in thin film conductors. Thin Solid Films 93:385
8. Rzepka S, Meusel E, Korhonen MA, Li C-Y (1999) 3-D finite element simulator for migration effects due to various driving forces in interconnect lines. In: AIP (ed) Stress-induced phenomena in metallization: fifth international workshop, vol 491, pp 150–161
9. Fick A (1855) Ueber Diffusion. Pogg Ann Phys Chem 170(4. Reihe 94):59–86
10. Dalleau D, Weide-Zaage K (2001) Three-dimensional voids simulation in chip metallization structures: a contribution to reliability evaluation. Microelectron Reliab 41:1625–1630
11. Weide-Zaage K, Dalleau D, Danto Y, Fremont H (2007) Dynamic void formation in a DD-copper-structure with different metallization geometry. Microelectron Reliab 47:319
12. Dalleau D, Weide-Zaage K, Danto Y (2003) Simulation of time depending void formation in copper, aluminum and tungsten plugged via structures. Microelectron Reliab 43:1821
13. Tan CM, Roy A (2006) Investigation of the effect of temperature and stress gradients on accelerated EM test for Cu narrow interconnects. Thin Solid Films 504:288
14. Tan CM, Li W, Tan KT, Low F (2006) Development of highly accelerated electromigration test. Microelectron Reliab 46:1638
15. Tan CM, Hou Y, Li W (2007) Revisit to the finite element modeling of electromigration for narrow interconnects. J Appl Phys 102:033705
16. Tan CM, Roy A (2007) Electromigration in ULSI interconnects. Mater Sci Eng Rev 58:1–75
17. Li W, Tan CM (2007) Enhanced finite element modelling of Cu electromigration using ANSYS and matlab. Microelectron Reliab 47:1497–1501

18. Shen Y-L, Ramamurty U (2003) Temperature-dependent inelastic response of passivated copper films: experiments, analyses, and implications. J Vac Sci Technol B 21:1258–1264
19. Tan CM, Roy A, Vairagar AV, Krishnamoorthy A, Mhaisalkar SG (2005) Current crowding effect on copper dual damascene via bottom failure for ULSI applications. IEEE Trans Dev Mater Reliab 5(2):198
20. Pyun JW, Baek W-C, Im J, Ho PS, Smith L, Neuman K, Pfeiler K (2006) Effect of barrier process on electromigration reliability of Cu/porous low-k interconnects. J Appl Phys 100:023532
21. Arnaud L, Tartavel G, Gerger T, Mariolle D, Gobil Y, Touet I (2000) Microstructure and electromigration in copper damascene lines. Microelectron Reliab 40:77
22. Glasow AV, Fischer AH, Steinlesberger G (2003) Using the temperature coefficient of the resistance (TCR) as early reliability indicator for stress voiding risks in Cu interconnects. In: IEEE 41st annual international reliability physics symposium (IRPS) proceedings
23. Lloyd JR, Clemens JJ, Snede S (1999) Copper metallization reliability. Microelectron Reliab 39:1595–1602
24. Shiley CG (1985) Steady-state temperature profiles in narrow thin-film conductors. J Appl Phys 57:777–784
25. Niwa H, Yagi H, Tsuchikawa H, Masaharu K (1990) Stress distribution in an aluminum interconnect of very large scale integration. J Appl Phys 68:328–333
26. Kreyszig E (1993) Advanced engineering mathematics, 7th edn. Wiley, New York
27. Gan ZH, Shao W, Mhaisalkar SG, Chen Z, Li H, Tu KN, Gusak AM (2006) Reservoir effect and the role of low current density regions on electromigration lifetimes in copper interconnects. J Mater Res 21:2241–2245
28. Shao W, Vairagar AV, Tung CH, Xie ZL, Krishnamoorthy A, Mhaisalkar SG (2005) Electromigration in copper damascene interconnects: reservoir effects and failure analysis. Surf Coat Technol 198:257–261
29. Duan QF, Shen Y-L (2000) On the prediction of electromigration voiding using stress-based modeling. J Appl Phys 87:4039–4041
30. Ogawa ET, Lee K-D, Matsuhashi H, Ko K-S, Justison PR, Ramamurthi AN, Bierwag AJ, Ho PS, Blaschke VA, Havemann RH (2001) Statistics of electromigration early failures in Cu/oxide dual-damascene interconnects. In Proceedings of the Conference 39th IEEE/IRPS, ed, Orlando, FL, USA, pp 341–349
31. Fischer AH, Glasow AV, Penka S, Ungar F (2002) Electromigration failure mechanism studies on copper interconnects. In: Interconnect technology conference, proceedings of the IEEE 2002 international, ed, pp 139–141
32. Ang D, Ramanujan RV (2006) Hydrostatic stress and hydrostatic stress gradients in passivated copper interconnects. Mater Sci Eng A 423:157–165
33. Vairagar AV, Mhaisalkar SG, Meyer MA, Zschech E, Krishnamoorthy A (2005) Reservoir effect on electromigration mechanisms in dual-damascene Cu interconnect structures. Microelectron Eng 82:675
34. Padhi D, Dixit G (2003) Effect of electron flow direction on model parameters of electromigration-induced failure of copper interconnects. J Appl Phys 94:6463–6467
35. Shen Y-L, Guo YL, Minor CA (2000) Voiding induced stress redistribution and its reliability implications in metal interconnects. Acta Mater 48:1667–1678
36. Vairagar AV, Mhaisalkar SG, Krishnamoorthy A (2004) Microelectron Reliab 44(5):747
37. Fu CM, Tan CM, Wu SH, Yao HB (2010) Width dependence of the effectiveness of reservoir length in improving electromigration for Cu/low-k interconnects. Microelectronics Reliab 50(9–11):1332–1335
38. Korhonen MA, Black RD, Li C-Y (1993) Stress evolution due to electromigration in confined metal lines. J Appl Phys 73:3790–3799
39. Tan CM, Raghavan N, Roy A (2007) Application of gamma distribution in electromigration for submicron interconnects. J Appl Phys 102:103703
40. Smy TJ, Winterton SS, Brett MJ (1993) A Monte Carlo computer simulation of electromigration. J Appl Phys 73:2821

41. Dew SK, Smy T, Brett MJ (1992) Simulation of elevated temperature aluminum metallization using SIMBAD. IEEE Trans Electron Dev 39:1599
42. Tait RN, Dew SK, Smy T, Brett MJ (1990) Ballistic simulation of optical coatings deposited over topography. In: SPIE international symposium on modeling of optical thin films II, Bellingham, WA, p 112
43. Lozowski EP, Brett MJ, Tait RN, Smy T (1991) Simulating giant hailstone structure with a ballistic aggregation model. Q J R Meteorol Soc 117:427
44. Bruschi P, Nannini A, Piotto M (2000) Three-dimensional Monte Carlo simulations of electromigration in polycrystalline thin films. Comput Mater Sci 17:299
45. Bruschi P, Cagoni P, Nannini A (1997) Temperature-dependent Monte Carlo simulations of thin metal film growth and percolation. Phys Rev B 55:7955
46. Amar JG, Family F, Amar G (1996) Kinetics of submonolayer and multilayer epitaxial growth. Thin Solid Films 272:208
47. Vairagar AV, Krishnamoorthy A, Tu KN, Mhaisalkar SG, Gusak AM, Meyer MA (2004) In situ observation of electromigration-induced void migration in dual-damascene Cu interconnect structures. Appl Phys Lett 85:2502
48. Zaporozhets TV, Gusak AM, Tu KN, Mhaisalka SG (2005) Three-dimensional simulation of void migration at the interface between thin metallic film and dielectric under electromigration. Appl Phys Lett 98:103508
49. Roy A, Tan CM, Kumar R, Chen XT (2005) Effect of test condition and stress free temperature on the electromigration failure of Cu dual damascene submicron interconnect line-via test structures. Microelectron Reliab 45:1443
50. Greenabaum B, Sauter AI, Flinn PA, Nix WD (1991) Stress in metal lines under passivation; comparison of experiment with finite element calculations. Appl Phys Lett 58:1845
51. Kilijanski MS, Shen Y-L (2002) Analysis of thermal stresses in metal interconnects with multilevel structures. Microelectron Reliab 42:259
52. Park Y-B, Jeon IS (2003) Mechanical stress evolution in metal interconnects for various line aspect ratios and passivation dielectrics. Microelectronics Eng 69:26
53. Huang JS, Yeh ECC, Zhang ZB, Tu KN (2002) The effect of contact resistance on current crowding and electromigration in ULSI multi-level interconnects. Material Chem Phys 77:377
54. Atkinson RR (2003) The State University of New Jersey, Ph.D. thesis, New Brunswick, Rutgers
55. Liu LM, Wang SQ, Ye HQ (2004) First-principles study of polar Al/TiN(1 1 1) interfaces. Acta Mater 52:3681
56. Smith JR, Zhang W (2000) The connection between ab initio calculations and interface adhesion measurements on metal/oxide systems: Ni/Al_2O_3 and Cu/Al_2O_3. Acta Mater 48:4395
57. ANSYS, Theory reference and reference therein
58. Guerard BV, Peisl H, Zitzmann R (1974) Appl Phys B 3:37
59. Carling KM, Wahnstrom G, Mattsson TR, Sandberg N, Grimvall G (2003) Vacancy concentration in Al from combined first-principles and model potential calculations. Phys Rev B 67:054101
60. Prybyla JA, Riege SP, Grabowski SP, Hunt AW (1998) Temperature dependence of electromigration dynamics in Al interconnects by real-time microscopy. Appl Phys Lett 73:1083
61. Shao W (2006) Investigation of surface and microstructure effect on electromigration of dual damascene Cu interconnects. Nanyang Technological University
62. Li W, Tan CM, Hou Y (2007) Dynamic simulation of electromigration in polycrystalline interconnect thin film using combined Monte Carlo algorithm and finite element modeling. J Appl Phys 101:104314

Chapter 5
Finite Element Method for Stress-Induced Voiding

5.1 Introduction

With the basic physics of stress-induced voiding (SIV) introduced in Chap. 2, the detailed finite element modeling of the mechanisms of SIV in Cu interconnect will be described here. The understanding of the voiding mechanism through the modeling can certainly shed light on the future design and process improvement of the multilevel interconnect structures. In three-dimensional (3D) modeling, the responses of interest are the normal stress components σ_x (along line length), σ_y (transverse to line), and σ_z (normal to line length); hydrostatic stress, von-Mises stress, and the corresponding stress gradient. Let us now look at the Finite Element Method (FEM) modeling for SIV.

5.2 FEM Model Description

Figure 5.1a shows a typical via–line structure with M1–V1–M2 configuration for the FEM model. The model geometry also includes M1 and M2 extensions (via-to-line overlap), which are widely used in practical layout. The unit of the dimensions shown in Fig. 5.1a is nanometer. The thickness of the barrier layer, Ta (tantalum) is 25 nm for the bottom and the side wall, and that of the etch stop layer USG (undoped silicate glass) is 50 nm. The height and width of the via are 280 and 260 nm, respectively. The entire line length of the model is fixed at 2,500 nm.

Figure 5.1b shows the finite element mesh with the boundary conditions for the via–line structure as given in Fig. 5.1a. As planes $x = 0, l$ and $y = 0, w$ are mirror symmetrical due to the repetitive structure, $u_x = 0$ in planes $x = 0$ and $x = l$; and $u_y = 0$ in planes $y = 0$ and $y = w$. Here, u_x and u_y correspond to the displacements along the x- and y-directions, respectively. The element type used in this

C. M. Tan et al., *Applications of Finite Element Methods for Reliability Studies on ULSI Interconnections*, Springer Series in Reliability Engineering, DOI: 10.1007/978-0-85729-310-7_5, © Springer-Verlag London Limited 2011

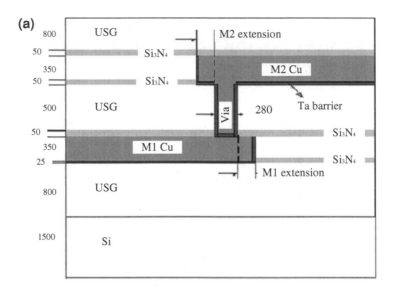

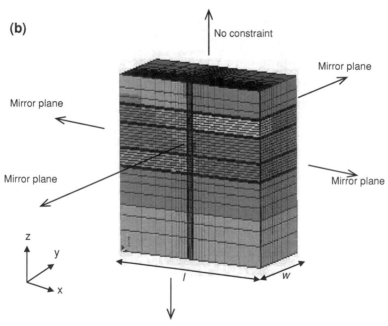

Fig. 5.1 **a** Model used for the via–line structure with geometry description and **b** model used for the via–line structure with the description of boundary condition of finite element mesh

ANSYS simulation is 8-node SOLID45, which is commonly used for 3D thermo-mechanical stress analysis [1]. A mesh refinement is adapted in the copper interconnects at those locations where high stress concentration is expected. Figure 5.2 gives a flow chart on using ANSYS for SIV modeling.

Table 5.1 lists the properties of the materials used in the simulation. These properties are obtained from the measurement of thin film since the model dimension is in the scale of micrometer. Each material is assumed to be an isotropic linear elastic solid without accounting for microstructure-dependent non-linear behavior of Cu. This assumption is reasonable because the width of the Cu lines is in the sub-micron range, and it continues to display linear elastic behavior even at temperatures as high as 400°C [2].

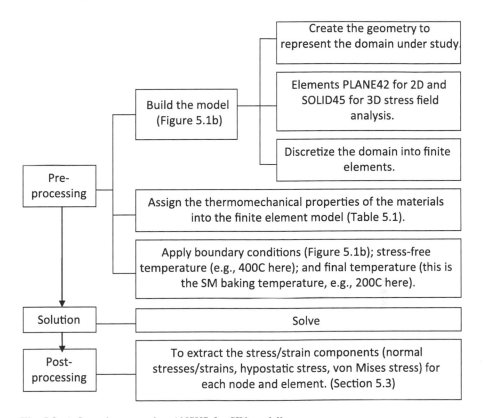

Fig. 5.2 A flow chart on using ANSYS for SIV modeling

Table 5.1 Thermo-mechanical properties of the materials used in the finite element analysis [3–9]

Properties	Materials	CTE (ppm/°C)	Modulus (GPa)	Poisson's ratio
Substrate	Si	2.6	130	0.28
IMD	USG	1.37	60	0.25
	CDO	12.0	16.2	0.25
	TEOS	1	59	0.16
	SiLK	62	3.5	0.35
Etch stop	SiN	3.2	220.8	0.27
Line	Cu	17.7	104.2	0.352
Barrier	Ta	6.5	185.7	0.342

Fig. 5.3 Normalized change in resistance after testing for 1,344 h at various test temperatures. Reprinted from Gan et al. [10]; copyright © 2006 with permission from Elsevier

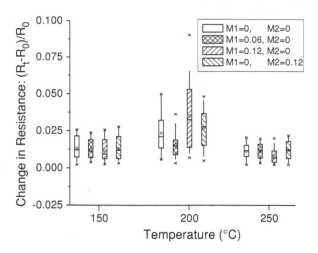

Thermo-mechanical calculations are performed for interconnect cooling from its initial stress-free temperature to the SIV test temperature of 200°C [10]. The stress-free temperature of 400°C is generally used since it is the deposition temperature of the dielectric material when the structure is supposed to be at the zero stress state [11]. The final temperature of 200°C is selected to ensure the highest acceleration in SIV and this temperature is termed as the critical temperature. The resistance changes of the Cu interconnects with varied M1 or M2 extensions in USG dielectric after 1,344 h of testing are shown in terms of test temperatures in Fig. 5.3. Since no bias is applied during testing, the observed resistance change and subsequent damage formation can be attributed directly to SIV.

From Fig. 5.3, the maximum resistance change with the widest spread is observed at 200°C. This observation is also reported in other literatures [12, 13]. Resistance changes observed at 150 and 250°C are similar to each other.

5.3 Finite Element Modeling Results

Different kinds of stress distribution in the copper interconnect such as von-Mises stress, hydrostatic stress, and principle stresses can be extracted from the FEM. As hydrostatic stress is known to be the driving force for SIV void nucleation [14], the hydrostatic stress distribution which is obtained from the simulation is able to predict the most probable void nucleation locations. The von-Mises stress, which is usually used as a criterion for plastic deformation, is in a form of octahedral shear stress which does not affect the volumetric changes of material. The hydrostatic stress (σ_{HYD}) and von-Mises stress (σ_Y) can be computed using Eqs. 5.1 and 5.2 as follows [6, 15]:

$$\sigma_{\mathrm{HYD}} = \frac{(\sigma_x + \sigma_y + \sigma_z)}{3}. \tag{5.1}$$

$$\sigma_{\mathrm{von-Mises}} = \frac{1}{\sqrt{2}}(\sigma_x - \sigma_y)^2 + (\sigma_y - \sigma_z)^2 + (\sigma_z - \sigma_x)^2 + 6\left(\tau_{xy}^2 + \tau_{yz}^2 + \tau_{xz}^2\right)\Big]^{\frac{1}{2}}. \tag{5.2}$$

Hydrostatic and von-Mises stress distribution of a via–line structure with TEOS dielectric are shown in Fig. 5.4. The highest and the second-highest hydrostatic stresses are found to be at the bottom plane of the via and the top plane of the lines, respectively, with their values around 600–700 MPa. Although a hydrostatic tensile stress of the order of 1 GPa would normally be needed to overcome

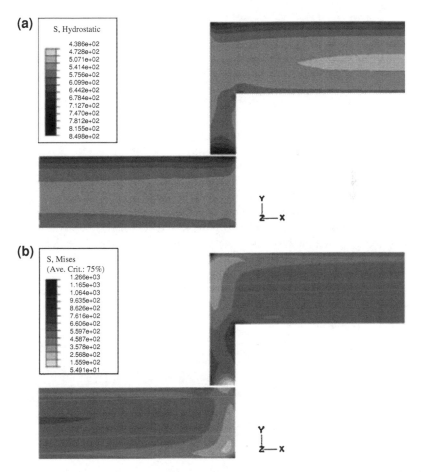

Fig. 5.4 Contour plots of **a** hydrostatic and **b** von-Mises stress of TEOS dielectric embedded via–line structure. Reprinted from Paik et al. [6], copyright © 2004 with permission from Elsevier

the energy barrier for void nucleation [16], it can be lowered when defects are present [17]. Thus, the bottom of the via can constitute a weak interface, unless the barrier deposition process is optimized. The top surface of the metal lines is likely to be contaminated or damaged during the chemical mechanical polishing (CMP) process and have a lot of defects and/or high interfacial energy. In other words, either the bottom of the via or the top of the lines will be the dominant nucleation sites due to the high density of the interfacial defects as well as high stress values in these regions, which agrees well with the experiments [18].

While the hydrostatic stress itself can explain the void nucleation sites, we found that it cannot well explain the SIV behavior. For example, it was reported that the simulated hydrostatic stress decreases with increasing line width [18]. However, experimental results indicated that wider metal lines have smaller mean time to failure, i.e., they are more susceptible to SIV when compared to narrower lines [19]. This phenomenon can be explained by the stress gradient concept instead.

In fact, once a void is nucleated, the stress gradient is believed to be the driving force for vacancy and/or atomic diffusion [20], as described in Eq. 5.3, causing the void to grow.

$$J_{SIV} = \pm \frac{D}{kT}\left(\frac{\Delta\sigma}{\Delta x}\right). \tag{5.3}$$

The negative and positive signs in Eq. 5.3 indicates vacancy and atomic fluxes, respectively.

Figure 5.5 gives a schematic of the situation when a Cu atom is subjected to a higher tensile stress at the right-hand side. We can imagine that the Cu atom will be driven toward the right direction (the side with higher tensile). In contrast, the vacancies will diffuse toward the more compressive (or lower tensile) region and coalescence or sink into voids where the stress will be relaxed. Voids can grow by the diffusion and accumulation of vacancies, until the stress gradient is eliminated. When this happens, the interconnect will experience severe reliability problems, such as open circuit failures.

In order to verify the accuracy of FEM in SIV modeling, let us apply the FEM to study the effects of various factors on SIV and compare the FEM results with actual experimental results.

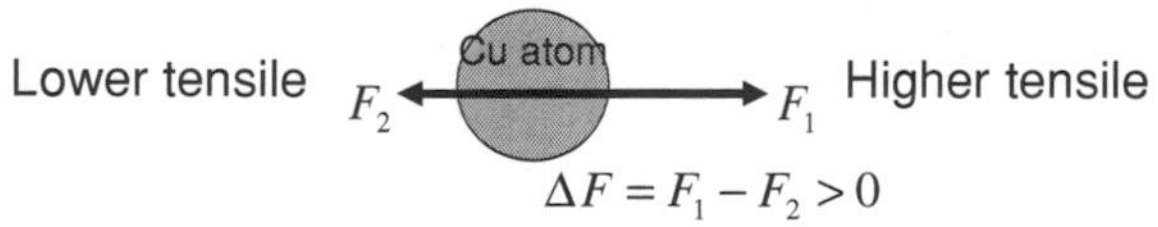

Fig. 5.5 Schematic for the situation when a Cu atom is subjected to a higher tensile stress at the right-hand-side

Lower tensile $\quad F_2 \longleftarrow$ Cu atom $\longrightarrow F_1$ Higher tensile

$$\Delta F = F_1 - F_2 > 0$$

5.4 Effect of Dielectric Materials on SIV

The effect of dielectric materials on SIV is studied using FEM, and the results are compared with reported experimental data. Two different dielectrics, USG and carbon doped oxide (CDO) are compared [10]. The changes in resistance of interconnects with these two dielectric materials are tested at 200°C, and their results are presented in Fig. 5.6 which shows a difference of two orders of magnitude in the resistance change between the interconnect structures with CDO and USG dielectrics. About 40% of the CDO samples showed open circuit failures after the 1,344 h of test, whereas the maximum resistance change in the USG samples was only 10%. The data clearly indicate that the choice of dielectric materials has a very significant impact on stress migration lifetimes.

Failure analysis using FIB indicated that failures in both CDO and USG were very similar in nature with voids nucleating symmetrically at the bottom of the via, and there are basically two modes of void growth, subsequently. In Mode I, the stress-induced voids grow vertically (Fig. 5.7a). In Mode II, voids prefer to grow horizontally along the via bottom (Fig. 5.7b). These two modes of void growth around the via bottom are schematically shown in Fig. 5.8.

As discussed earlier, the site of the most concentrated hydrostatic tensile stress is believed to be the most probable void nucleation site. Regardless of the dielectrics employed, the simulated concentration center of hydrostatic tensile stress is indeed found situated at the edge of the via as shown in Fig. 5.9, in agreement with the actual failure behaviors of the via–line structure in Fig. 5.7.

Fig. 5.6 Changes in resistance of USG vs CDO ILD after 1,344 h, test temperature = 200°C. Reprinted from Gan et al. [10], copyright © 2006 with permission from Elsevier

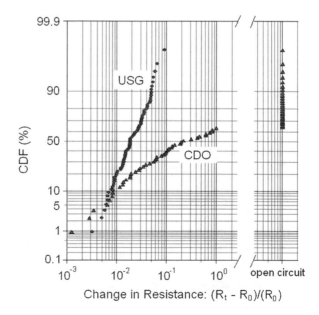

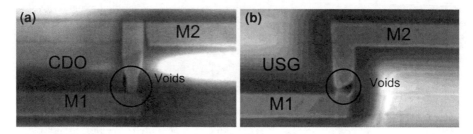

Fig. 5.7 Post-failure FIB images. **a** Mode I, dielectric = CDO; **b** Mode II, dielectric = USG. Reprinted from Gan et al. [10], copyright © 2006 with permission from Elsevier

Fig. 5.8 Schematic of two modes of void formation at the via bottom: **a** voids stretch *vertically* and **b** void grows *horizontally*, leading to a more detrimental situation. Reprinted from Gan et al. [10], copyright © 2006 with permission from Elsevier

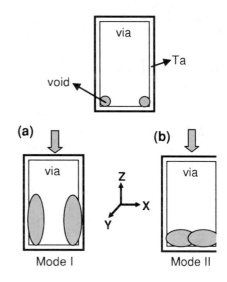

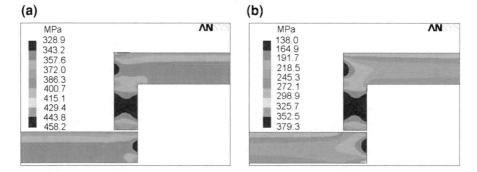

Fig. 5.9 Finite element simulated contour plot of hydrostatic stress around via bottom **a** for USG and **b** for CDO dielectrics. Reprinted from Gan et al. [10], copyright © 2006 with permission from Elsevier

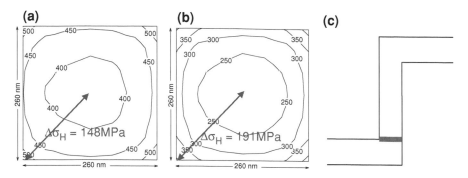

Fig. 5.10 Hydrostatic stress contours at via bottom for **a** USG dielectric and **b** CDO dielectric; **c** schematic cross-section of the metal–via structure highlighting the via bottom

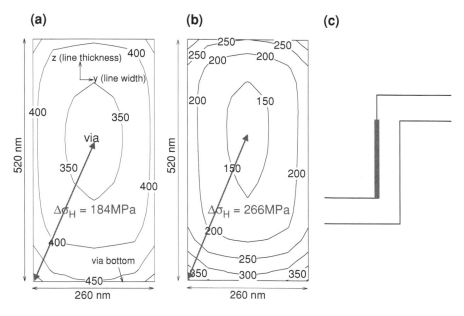

Fig. 5.11 Hydrostatic stress contours at via side wall for **a** USG dielectric and **b** CDO dielectric; **c** schematic cross-section of the metal-via structure highlighting the via side wall

The inclusion of M1 or M2 extension does not have obvious effect on the stress distribution. Indeed, the via bottom is a process-related weak interface that lowers the SIV reliability especially in narrow interconnects. The crack propagates rapidly due to large stress concentration at the void tip.

To explain the experimental results of smaller SIV lifetime for interconnects with CDO dielectric, FEM is employed and the hydrostatic stress gradients are computed. The results are shown in Figs. 5.10 and 5.11. As shown in Fig. 5.10, the average stress gradient at the via bottom in CDO structure along the diagonal on the plane of via bottom is about 1.039 MPa/nm, which is about 30% higher than that in USG structure (~ 0.805 MPa/nm). Similarly, as shown in Fig. 5.11, the average stress gradient at the via side wall in CDO structure along the z-axis (~ 0.578 MPa/nm) is about 57% higher than that in USG structure (~ 0.368 MPa/nm). As higher stress gradient in CDO structure will lead to higher driving force for void growth, the calculated higher stress gradient in CDO could be one of the key reasons for its higher failure rate.

Besides the hydrostatic stress and stress gradient, which affect the void formation and growth, other factors, such as the back stress during stress migration and interface strength between Cu lines and the dielectrics, may also affect the void growth. Discussion of the two factors is given below.

Similar to the electromigration, back stress may also be established during SIV. This back stress serves as resistance to the stress-induced voiding. By considering the back stress effect, the net atomic flux due to SIV can be expressed as follows:

$$J_{SM} = \frac{D}{kT}\left(\frac{d\sigma_H}{dx} - \frac{d\sigma_{BF}}{dx}\right). \tag{5.4}$$

Here $d\sigma_{BF}/dx$ is the back stress gradient along the line. Qualitatively, the magnitude of back stress gradient depends on the confinement provided by the elastic property of the dielectric and the metal/dielectric interfacial adhesion [21]. Stronger confinement leads to higher back stress gradient. From Table 5.1, Young's modulus of CDO is about ¼ of USG. If the adhesion is strong, the high elasticity modulus of USG provides a stronger confinement to the Cu lines. Indeed, experimental measurement by Webb et al. [22] showed that the adhesion strength of Ta/low-k interface was generally lower (<10 J/m^2) than that of Ta/SiO$_2$ interface (>10 J/m^2). Therefore, it is expected that the combined effect of strong confinement and good adhesion in the USG samples will lead to higher back stress in the samples. It is believed that the stronger back stress effect in USG samples, help in retarding the voiding when compared to its CDO counterpart.

To simulate the void growing process after its nucleation, FEM calculation is performed with the aid of element removal technique, which has been implemented in the dynamic simulation of EM described in Chap. 3. After thermomechanical analysis, the element corresponding to the maximum hydrostatic stress gradient is deleted to simulate the voiding process. The simulation and element deletion are repeated over and over again, to simulate the void growing process.

Figure 5.12 shows the simulated voiding processes, and the processes are consistent with the two failure modes observed in Fig. 5.7. Figure 5.12a–c shows the voiding process for USG sample while Fig. 5.12d–f shows the voiding process for CDO sample. In both cases, voids are observed to start from the outer corner of the via bottom. After that, void grows horizontally along via bottom for USG interconnects as shown in Fig. 5.12a–c. For CDO interconnects, void grows along the inner via sidewall as shown in Fig. 5.12d–f, in agreement with the experimental observations.

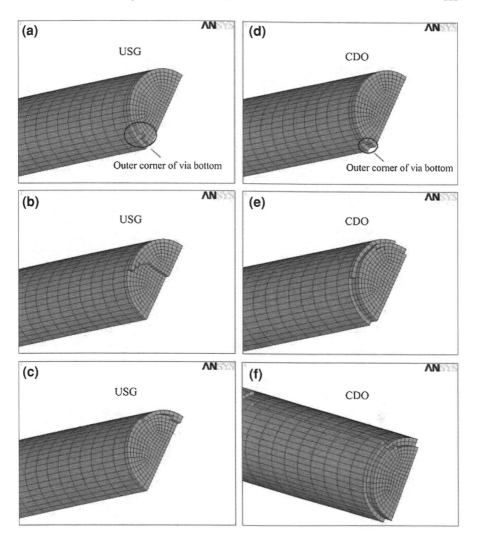

Fig. 5.12 Two different failure modes for SIV failure in via simulated by ANSYS$^{@}$. **a–c** CDO interconnect; **d–f** USG interconnect. Reprinted from Hou and Tan [23], copyright © 2009 with permission from Elsevier

5.5   Effect of Lower Layer Line Width on SIV

The normalized changes in resistance of M1 test structures with various line widths are shown in Fig. 5.13, assuming a lognormal distribution [24]. It can be clearly seen from Fig. 5.13 that SIV reliability is better for narrower M1 lines. However, the change in resistance may not have much difference when the line width of M1 increases from 700 to 1000 nm as they have similar resistance change trend as can be seen in Fig. 5.13.

Fig. 5.13 Normalized change in resistance of M1 Cu line for various line widths. Reprinted from Shao et al. [24], copyright © 2006 with permission from Elsevier

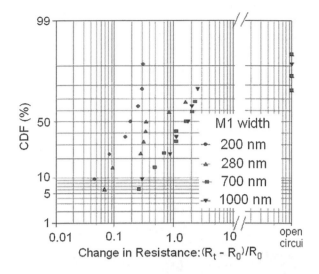

Fig. 5.14 TEM of a stress-induced void under the via in the wide metal line. Reprinted from Yoshida et al. [25], copyright © 2002, IEEE

The SIV failure mode of the structure with via above a wide metal was generally related to the metal surface voiding underneath the via as shown in Fig. 5.14. Cross sectional FIB micrographs around Cu via of the test structures with 700 nm M1 line width after 30% resistance increase is shown in Fig. 5.15. It can be seen from Fig. 5.15 that the voiding took place at the M1 Cu/dielectric interface, away from the via bottom.

Figure 5.16 shows the hydrostatic stress distribution in the via-line structure for M1 line width of 700 nm, and one can observe that hydrostatic tensile stress is not uniform throughout the Cu structure. The highest and the second-highest hydrostatic tensile stress are located at the bottom plane of the via and at the top plane of

Fig. 5.15 SIV failure modes in M1 with line width of 700 nm. Reprinted from Shao et al. [24], copyright © 2006 with permission from Elsevier

Fig. 5.16 Hydrostatic stress distribution in via–line structure for M1 line width of 700 nm. Reprinted from Shao et al. [24], copyright © 2006 with permission from Elsevier

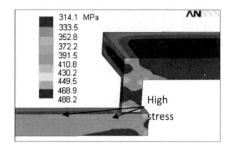

M1 Cu/SiN interface, respectively. This indicates that void is highly probable to nucleate at the top surface of M1 and at the bottom of the via.

However, no void is found at the bottom of the via even though it is a highly stress concentrated site. This could be explained by the existence of the diffusion barrier layer which cuts off the vacancy diffusion path around the via. Earlier studies have shown that the interface region between copper and Ta has much better adhesion when compared to the interface region between copper and Si_3N_4 [26]. The strong interface between copper and Ta prevents vacancies from diffusing through the barrier layer and into the bottom of the via. Therefore, the void is constrained to form only at the top surface of M1.

As the stress gradient acts as the driving force for vacancy movement, it is computed and plotted along the M1Cu/Si_3N_4 interface as shown in Fig. 5.17. It can be seen that the stress gradient ($grad\sigma_l$) along the length direction is not a strong function of line width when compared to that along the width direction ($grad\sigma_w$), and the gradient increases as the line width increases. Therefore, the stress gradient along the width direction is believed to play an important role in influencing the rate of vacancy diffusion, and this can be explained as follows.

A schematic diagram is shown in Fig. 5.18. With the co-existence of the stress gradients along the width and length directions, the resultant stress gradient can be found using the vector sum and the direction is denoted as R in Fig. 5.18. Since the driving force along the width direction is higher (due to larger stress gradient) in wider line, the resultant direction R is steeper, which indicates that the vacancies

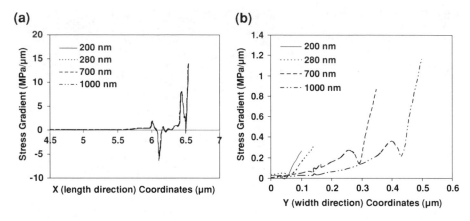

Fig. 5.17 Variation of the hydrostatic stress gradient for different M1 line widths: **a** stress gradient along the length direction (*x*-direction) gradσ_l at the M1/cap interface and at $y = 0$; **b** stress gradient along the width direction (*y*-direction) gradσ_w at the M1/cap interface and $x = 5$ μm. Reprinted from Shao et al. [24], copyright © 2006 with permission from Elsevier

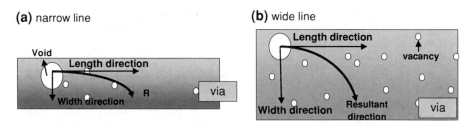

Fig. 5.18 Schematic diagram showing that larger stress gradient along the width direction in *wide line* (**b**) will lead to the vacancy driving force in a steeper direction when compared to *narrow line* (**a**). Reprinted from Shao et al. [24], copyright © 2006 with permission from Elsevier

Fig. 5.19 Volume averaged stress gradients versus line width. Reprinted from Shao et al. [24], copyright © 2006 with permission from Elsevier

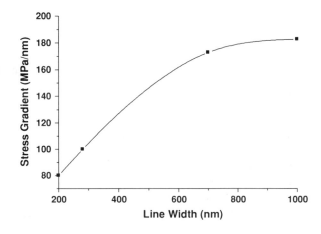

diffuse to the line edge beneath the via bottom at a much faster rate, rendering a higher resistance change when compared to its narrower line counterpart.

As higher stress gradients signify an increase in the driving force for vacancy movement, SIV failure times can be defined by the amount of stress gradient increase. In order to seek a correlation between SIV reliability and line width, the volume-averaged stress gradient is plotted against line width as shown in Fig. 5.19. It is clear from the figure that the magnitude of the stress gradient increases with line width, indicating that wider lines are more prone to void growth, and thus a shorter failure time. However, when the line width becomes larger than 700 nm, the amount of driving force increases at a decreasing rate, resulting in a saturation of the medium time to failure, in agreement with the experimental data reported earlier. The FEM simulation suggested that the difference in the stress gradients is the root cause of the line width effect.

5.6   Effect of via-Misalignment on SM

Via-to-line misalignment was intentionally introduced through specially designed test structures [5]. The via-to-line overlap varied from -0.06 to $+0.2$ μm in both metal levels separately (Fig. 5.20), and the corresponding failure time distributions with varying overlap between W-plug and copper level are summarized in Fig. 5.21. It can be seen that the distributions have similar shape factors but failure times are steadily smaller with smaller overlap. The MTF is reduced by about 2X when the overlap decreases from 0.00 to -0.06 μm. On the other hand, the failure time is increased when the overlap increases from 0.00 to $+0.20$ μm, although the gain is less pronounced.

Figure 5.22 gave the FEM simulations of the stress components for via–line structures with and without misalignment under the elastic deformation assumption. The regions of high stress gradients are denoted by the alphabets A–E. High stress gradient for σ_{xx} is located at the upper part of the metal line next to the via,

Fig. 5.20 Failure distributions obtained at 275°C on short via chains as a function of the Cu-to-W-via overlap using 5% relative failure criterion. The failure times become smaller with smaller overlap between copper line and W-plug. Reprinted with permission from Glasow et al. [18], copyright © 2002, AMC

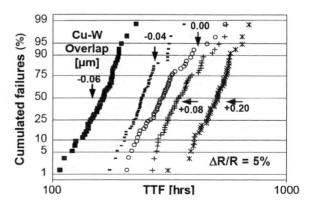

Fig. 5.21 MTFs obtained at 275°C on short via chains as a function of the Cu-W and Al-W overlap, respectively

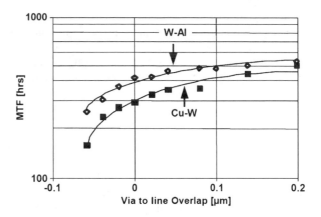

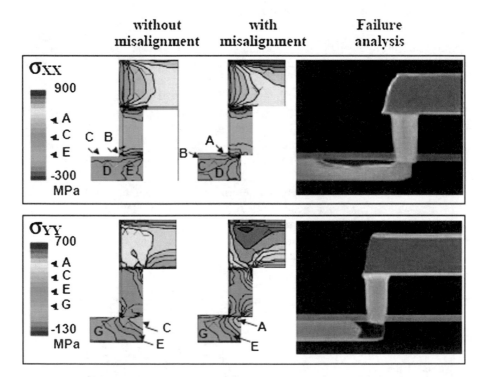

Fig. 5.22 FEM simulations of the linear σ_{xx} (*top*) and σ_{yy} (*bottom*) stress components for via–line structures without and with misalignment of the line below the via, respectively. Regions of high σ_{xx} stress gradients are obtained in the upper part of the metal line next to the via; σ_{yy} gradients directly under the via. In both regions stress-induced voids can be found after HTS test corresponding to two different SV failure modes. Reprinted with permission from Glasow et al. [18], copyright © 2002, AMC

whereas high σ_{yy} gradient was located directly under the via. The corresponding stress-induced voids were also shown in Fig. 5.22. They can be attributed to different failure modes with different activation energies [27].

Figure 5.22 also shows that the stress gradients in both x and y directions for sample with misalignment, is larger than the case without misalignment. Consequently, higher driving forces are expected at a large negative via-to-line overlap leading to a reduction of SIV failure times on those structures, as observed experimentally.

5.7 Conclusion

FEM provides insightful analysis to understand the SIV behavior for advanced interconnect systems. The method of FEM for SIV investigation is presented, and the effect of dielectric materials, lower layer line width, and via-misalignment on SIV are understood through the finite element analysis.

References

1. ANSYS Theory Reference (1999) Release 5.6. Swanson Analysis system, Inc. (now ANSYS Inc.), Canosburg, USA
2. Rhee SH, Du Y, Ho PS (2003) Thermal stress characteristics of Cu/oxide and Cu/low-k submicron interconnect structures. J Appl Phys 93:3926–3933
3. Tan CM, Gan ZH, Gao XF (2003) Temperature and stress distribution in the SOI structure during fabrication. IEEE Trans Semicond Manufactur 16:314–318
4. Zhao JH, Ryan T, Ho PS, Mckerrow AJ, Shih W-Y (1999) Measurement of elastic modulus, Poisson ratio, and coefficient of thermal expansion of on-wafer submicron films. J Appl Phys 85:6421–6424
5. Grill A (2003) Plasma enhanced chemical vapor deposited SiCOH dielectrics: from low-k to extreme low-k interconnect materials. J Appl Phys 93:1785
6. Paik JM, Park H, Joo YC (2004) Effect of low-k dielectric on stress and stress-induced damage in Cu interconnects. Microelectron Eng 71:348–357
7. CRC Handbook of Materials Science (1975) vol II, CRC Press, Boca Raton
8. Zhao JH, Du Y, Morgen M, Ho PS (2000) Simultaneous measurement of Young's modulus, Poisson ratio, and coefficient of thermal expansion of thin films on substrates. J Appl Phys 87:1575–1577
9. Brandes EA (1999) Smithells metals reference book. 7th edn. Butterworth-Heinemann, Oxford
10. Gan ZH, Shao W, Mhaisalkar SG, Chen Z, Li HY (2006) The influence of temperature and dielectric materials on stress induced voiding in Cu dual damascene interconnects. Thin Solid Films 504:161–165
11. Shi LT, Tu KN (1994) Finite-element modeling of stress distribution and migration in interconnecting studs of a three-dimensional multilevel device structure. Appl Phys Lett 65:1516–1518
12. Ogawa ET, Mcpherson JW, Rosal JA (2002) Stress-induced voiding under vias connected to wide Cu metal leads In: 40th annual IEEE international reliability physics symposium (IRPS) Proceedings, pp 312–321

13. Li B, Sullivan TD, Lee TC, Badami D (2004) Reliability challenges for copper interconnect. Microelectron Reliab 44:365–380
14. Kawano M, Fukase T, Yamamoto Y (2003) Stress relaxation in dual-damascene Cu interconnects to suppress stress-induced voiding. In: IEEE international interconnect technology conference, pp 210–212
15. Dieter GE (2001) Elements of the theory of plasticity. McGraw-Hill, London
16. Nix WD, Arzt E (1992) On void nucleation and growth in metal interconnect lines under electromigration conditions. Metallur Mater Trans A 23:2007–2013
17. Clement BM, Nix WD, Gleixner RJ (1997) Void nucleation on a contaminated patch. J Mater Res 12:2038–2042
18. Glasow AV, Fischer AH, Hierlemann M, Penka S, Ungar F (2002) Geometrical aspects of stress-induced voiding in copper interconnects. In: Advanced metallization conference (AMC), pp 161–167
19. Park YB, Jeon IS (2004) Effects of mechanical stress at no current stressed area on electromigration reliability of multilevel interconnects. Microelectron Engineering 71:76–89
20. Okabayashi H (1993) Stress-induced void formation in metallization for integrated circuits. Mater Sci Eng 11:191–241
21. Lee KD, Ogawa ET, Yoon S, Lu X, Ho PS (2003) Electromigration reliability of dual-damascene Cu/porous methylsilsesquioxane low k interconnects. Appl Phys Lett 82:2032–2034
22. Webb E, Witt C, Andryuschenko T, Reid J (2004) Integration of thin electroless copper films in copper interconnect metallization. J Appl Electrochem 34:291–300
23. Hou Y, Tan CM (2009) Comparison of stress-induced voiding phenomena in copper line-via structures with different dielectric materials. Semicond Sci Technol 24:085014
24. Shao W, Gan ZH, Mhaisalkar SG, Chen Z, Li HY (2006) The effect of line width on stress-induced voiding in Cu dual damascene interconnects. Thin Solid Films 504:298–301
25. Yoshida K, Fujimaki T, Miyamoto K, Honma T, Kaneko H, Nakazawa H, Morita M (2002) Stress-induced voiding phenomena for an actual CMOS LSI interconnects. IEDM, IEEE, pp 753–756
26. Hu C-K, Gignac L, Liniger E, Herbst B, Rath DL, Chen ST, Kaldor S, Simon A, Tseng W-T (2003) Comparison of Cu electromigration lifetime in Cu interconnects coated with various caps. Appl Phys Lett 83:869–871
27. Glasow AV, Fischer AH (2002) New approaches for the assessment of stress-induced voiding in Cu interconnects. In: Proceedings of the IEEE International Interconnect Technology Conference (IITC) pp. 274–276

Chapter 6
Finite Element Method for Dielectric Reliability

6.1 FEM on *k*-Value Extraction for Low-*k* Material

The effective dielectric constant (K_{eff}) is a concept to characterize the integrated working permittivity of a structure consisting of various dielectrics of different dielectric constants. K_{eff} of a specific structure depends on the ratio and number of the different dielectric parts. The requirements of the ITRS 2005 roadmap concerning effective *k* are given in Table 6.1. Owing to the complex structure of dual damascene interconnection, this value is not practically measurable, it can, however, be determined using FEM simulation.

This section describes the basic procedure to extract the *k*-value of a low-*k* material from a comb interline structure as shown in Fig. 6.1, using the commercially available FEM software ANSYS. In order to extract *k*-value from this kind of complex structure, the FEM simulation is incorporated into the analysis that involves the capacitance measurement and the characterization of the actual dimensions of the dielectric spacing/dimensions from cross-sectional images (using FIB, for example). The procedure is shown in Fig. 6.2. In the FEM, an initial *k*-value of the low-*k* material under study is assumed as input for the model, which in turn will give a simulated capacitance of the interline structure. The simulated capacitance will then be compared to the measured value. If both capacitances fit well to each other, then the initially assumed *k*-value will be considered as the real effective *k*-value for the low-*k* dielectric under study. On the other hand, if the simulated capacitance does not fit the measured one, then a new *k*-value will be assumed, until a good fit is achieved. In general, several iterations are necessary.

ANSYS develops a special command macro (CMATRIX) [2] to extract self- and mutual-capacitance terms for multiple conductor systems. The element type available for 2D modelling of the structure is an eight-node electrostatic solid element PLANE121 [3]. Figure 6.3 shows a 2D finite element mesh and the electric field vector of a parallel plate capacitor for verification purpose.

C. M. Tan et al., *Applications of Finite Element Methods for Reliability Studies on ULSI Interconnections*, Springer Series in Reliability Engineering, DOI: 10.1007/978-0-85729-310-7_6, © Springer-Verlag London Limited 2011

Table 6.1 ITRS 2005 MPU interconnect technology requirements for the 65- and 45-nm node [1]

Year of production	2007	2010
Technology node (nm)	65	45
Effective dielectric constant for the interlevel metal insulator	2.7–3.0	2.5–2.8

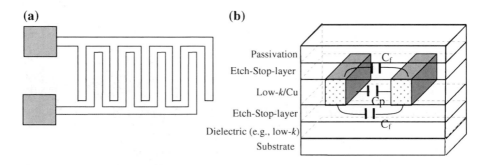

Fig. 6.1 a Schematic top view of the comb structure (interline capacitor); **b** Schematic cross-sectional view of the capacitance components involved in two metal fingers. The simplified capacitance components Cp and Cf refer to the parallel capacitance and fringing capacitance, respectively

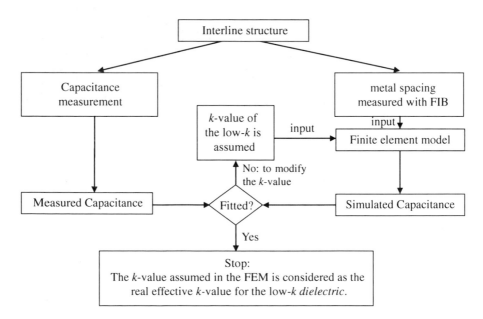

Fig. 6.2 k-value extraction methodology from interline capacitance structures based on physical and electrical measurement and FEM simulation

Fig. 6.3 A 2D finite element
mesh and the electrical field
vector of a parallel plate
capacitor for verification
purpose

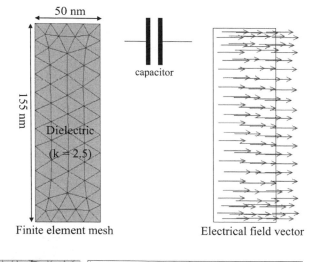

Finite element mesh   Electrical field vector

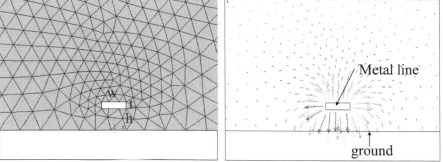

Fig. 6.4 A 2D finite element mesh and the electrical field vector of a metal above a ground plane
for verification purpose

The capacitance of the parallel plate capacitor from this numerical analysis is
0.068619 fF/um, which is very close to the analytical value of 0.0686185 fF/um as
computed below.

$$C = k \cdot \varepsilon_0 \frac{A}{d} \tag{6.1}$$

where d is the capacitor spacing, A is the capacitor area and ε_0 is the vacuum
permittivity.

Figure 6.4 shows a 2D finite element mesh and the electric field vector of a
metal above a ground plane for another verification purpose. Chang [4] reported
analytical formulas for the capacitance of a single metal-line over a ground plane
in a uniform dielectric given by Eqs. 6.2–6.8 as follows.

$$C = \frac{2\varepsilon}{\pi} \ln \frac{2R_b}{R_a} \tag{6.2}$$

$$\ln R_a = -1 - \frac{\pi w}{2h} - \frac{p+1}{\sqrt{p}} \tanh^{-1}\left(\frac{1}{\sqrt{p}}\right) - \ln\left(\frac{p-1}{4p}\right) \tag{6.3}$$

$$\ln R_b = \eta + \frac{p+1}{2\sqrt{p}} \ln \Delta \tag{6.4}$$

$$\eta = \sqrt{p}\left[\frac{\pi w}{2h} + \frac{p+1}{2\sqrt{p}}\left(1 + \ln\left(\frac{4}{p-1}\right)\right) - 2\tanh^{-1}\left(\frac{1}{\sqrt{p}}\right)\right] \tag{6.5}$$

$$\Delta = \text{larger value of } \eta \text{ or } p \tag{6.6}$$

$$p = 2B^2 - 1 + \sqrt{\left(2B^2 - 1\right)^2 - 1} \tag{6.7}$$

$$B = 1 + \frac{t}{h} \tag{6.8}$$

Table 6.2 compares the capacitance from the finite element simulation and that from the analytical formulas, indicating that the difference is less than 2%. It is noted that more accurate simulation can be achieved by a finer mesh which is a trade-off of a higher computational time.

Figure 6.5 shows the definition of a 2D model with two electrodes and one ground [5]. A configuration of a cross-sectional structure is given in Fig. 6.6 [5]. The basic idea for the extraction of the effective dielectric constants is to fit the simulated result to the measured capacitance by considering not only the dielectric

Table 6.2 The comparison of the capacitance from FEM and that from analytical formulas

w/h	t/h	C1 (=C/ε*l) Analytical	C2 (=C/ε*l) Numerical	(C2−C1)/C1*100%
1.12	0.318	3.552	3.620	1.92
2.01	0.485	4.762	4.840	1.63
2.52	0.318	5.227	5.256	0.56
2.72	0.802	5.789	5.866	1.32
3.18	0.936	6.371	6.451	1.25
3.63	0.802	6.806	6.893	1.28
3.68	0.485	6.662	6.692	0.45
3.79	1.116	7.136	7.216	1.12
4.24	0.936	7.542	7.595	0.70
5.44	1.2	8.961	9.000	0.44
6.36	0.802	9.760	9.760	0.00
7.42	0.936	10.955	10.947	−0.07
8.21	1.805	12.122	12.106	−0.13
9.85	1.805	13.842	13.788	−0.39
11.9	1.453	15.855	15.736	−0.75
14.78	1.805	18.951	18.779	−0.91
22.22	0.407	27.074	26.790	−1.05
58.25	7.113	63.940	63.141	−1.25

Fig. 6.5 The definition of
two-dimensional 2 electrodes
and 1 ground model (ANSYS
CMATRIX macro model)
[5], Copyright © 2005, IEEE

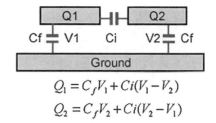

$$Q_1 = C_f V_1 + Ci(V_1 - V_2)$$
$$Q_2 = C_f V_2 + Ci(V_2 - V_1)$$

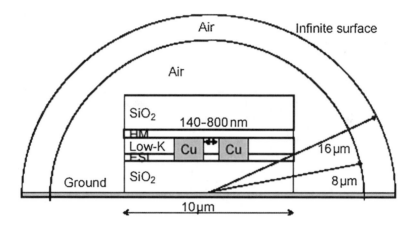

Fig. 6.6 The configuration of the parameters of the FEM simulation [5], Copyright © 2005, IEEE

under study but also the functional dielectric layers such as the etch-stop-layer (ESL) shown beneath the 'Low-K' in Fig. 6.6. The procedure is described in Fig. 6.2. Figure 6.7a, b shows the dependence of measured and simulated interline capacitances on Cu damascene spacing for low-*k* and SiO$_2$ films, respectively. If the low-*k* dielectric constant is fixed at 2.1, then the simulated capacitance fits well with the measured one when the spacing is 800 nm. However, the simulated capacitance becomes smaller and less than the measured result when the spacing continue to decrease as shown in Fig. 6.7a. This is believed to be related to the process-induced damage which will be discussed in detail in next section. On the other hand, Fig. 6.7b shows that there is no difference observed in the case where the dielectric under study is SiO$_2$, indicating that the SiO$_2$ is much less possibly resulted from process-induced damage.

6.2 FEM for Process-Induced Damage Evaluation

The interline capacitance will be changed due to process-induced damages, such as dry etching, cleaning, film deposition and CMP [6]. The discrepancy between the simulated and measured capacitance in Fig. 6.7a is because the dielectric constant

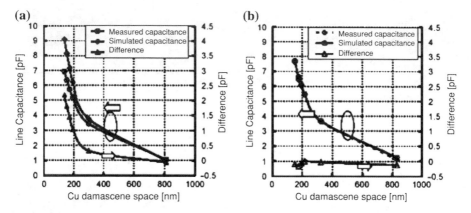

(a)   **(b)**

Fig. 6.7 Dependence of measured and simulated interlines capacitance on Cu damascene space for **a** low-k insulator; and **b** SiO$_2$ insulator [5], Copyright © 2005, IEEE

Fig. 6.8 Dependence of dielectric constant of low-k film on Cu damascene spacing [5], Copyright © 2005, IEEE

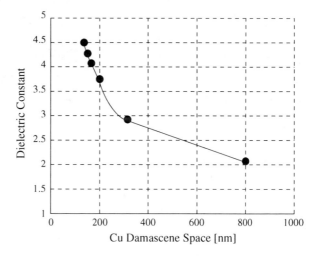

of the low-k is assumed to be constant in the simulation, which is not necessarily true when there is process-induced damage. Figure 6.7a indicates that the discrepancy in capacitance becomes in capacitance more aggressive when low-k is integrated and the metal spacing is reduced to less than 800 nm. The effective dielectric constant can be considered as being modified when the spacing is narrower. Figure 6.8 gives the dependence of the dielectric constant of low-k film on Cu damascene spacing by fitting the simulated result to the measured capacitance. The extracted dielectric constant of the low-k film increased from 2.1 to 4.5 for spacing from 800 to 140 nm.

Three possible damage modes, namely interface degradation, void-like degradation, and sidewall degradation were proposed by Chikaki et al. [5]. They studied the impacts of these degradations on capacitance changes using finite element simulation. Figure 6.9 shows the schematic cross-sectional diagram and the finite element mesh for these three damage modes.

Figure 6.10a shows that, when an interfacial damaged layer having a thickness of 10 nm and a k-value of 34.5 is added to the FEM model as shown in Fig. 6.9a, the simulated capacitance can fit very well with the measured capacitance. It is supposed that this interfacial damage layer may be formed because of wet chemicals, such as CMP, Cu electrochemical plating, and/or wet cleaning.

Figure 6.10b shows the comparison of the simulated interlines capacitance with those measured values, assuming that there is void formed in the low-k film. The simulated data fit well to the measured data if the void diameters ranged from 55 nm (with $k = 12.5$) to 47 nm (with $k = 80$). However, the k-value of a void is generally believed to be close to 1. Therefore, the k-value greater than 12 used here with the void-like degradation model as shown in Fig. 6.9b will be unlikely. In other words, void should not be the root cause for the damage.

If the side-wall degradation model as shown in Figs. 6.9c and 6.10c is assumed, and the thickness of the damaged layer ranges from 65 to 29 nm with k-values of 4.5–80, then a misfit value of less than 15% can be obtained. Such lateral sidewall damage was thought to be caused by dry etching and ashing. This is another plausible mechanism which contributes to the capacitance increase with the metal spacing reduction.

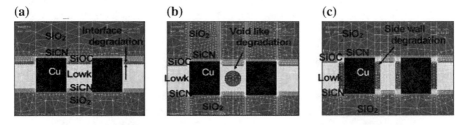

Fig. 6.9 Schematic cross section and the finite element mesh for three plausible damage modes: **a** Interface degradation mode; **b** Void-like degradation mode; and **c** Side-wall degradation mode [5], Copyright © 2005, IEEE

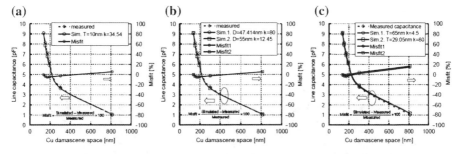

Fig. 6.10 Comparison between the measured and simulated capacitance in terms of Cu damascene spacing for **a** Interface degradation mode; **b** Void-like degradation mode and **c** Side-wall degradation mode [5], Copyright © 2005, IEEE

From the above description, we can see that it is not possible to identify the main mechanism for the degradation with this simple FEM analysis. A full analysis combined with failure analysis and interline leakage (V-ramp) evaluation is necessary.

6.3 FEM for *k*-Drift of Low-*k* Dielectric Materials

Some low-*k* dielectric materials can degrade with time due to chemical changes, moisture adsorption, vacuuming, and other reasons [7, 8], and the rate of the degradation in *k* can be accelerated by high temperatures. As a result, low-*k* dielectric materials show a drift in the measured dielectric constant as a function of time and temperature. This *k*-drift can affect the speed of an IC causing the device to fail some time in the future. This section describes a method to evaluate the *k*-drift of low-*k* dielectrics by combining the capacitance measurement and FEM. After *k*-drift is extracted as a function of time and temperature, it could be applied to the SPICE model to evaluate the impact of *k*-drift of the low-*k* dielectrics on the reliability and device performance.

Figure 6.11 shows the FIB cross section of the comb structure under study. The comb structure is shown in Fig. 6.1. The corresponding FEM is shown in Fig. 6.12, including the 2D finite element mesh and the electric field vector obtained from ANSYS.

Figure 6.13 gives the simulated capacitance in terms of *k* values of the low-*k* and the finger spacing, where the finger width is fixed at 120 nm. It is clear that the capacitance is linearly dependent on the *k* values of the low-*k*. This fitting methodology is similar to the previous report by Choudhury and Sangiovanni-Vincentelli [9]. The gradient *dC/dk* is also plotted in terms of finger spacing and finger width in Fig. 6.14, which fit nicely to a function given in Eq. 6.9:

$$\frac{dC}{dk} = 0.010447 + \frac{2.19}{\text{space}} - \frac{8.293}{\text{space}^2} - \frac{0.916}{\text{space} + \text{width}} \tag{6.9}$$

where space and width are in nm; *dC/dk* in fF/um. Hence, by measuring ΔC due to degradation, it is straightforward to compare the *k*-drift (Δk) given that the spacing and width are known.

Fig. 6.11 The FIB cross section of the comb structure, indicating the spacing is varied from 4.7 to 5.2 nm

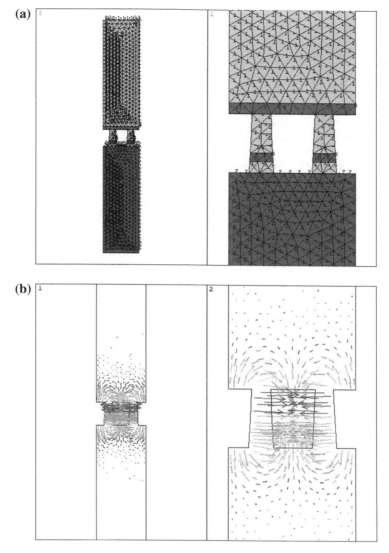

Fig. 6.12 **a** A two-dimensional finite element mesh and **b** the electrical field vector of the comb structure according to the dimensions from Fig. 6.11 including two electrodes only

6.4 FEM for Electric Field Simulation

ANSYS provides both 2D- and 3D elements for electric field analysis. PLANE121 is a 2D, eight-node, charge-based electrical element (Fig. 6.15). The element has one degree of freedom (i.e. voltage) at each node. The eight-node elements have compatible voltage shapes and are well suited to model curved boundaries.

Fig. 6.13 The calculated capacitance in terms of k of low-k and the finger spacing, where the finger width is 120 nm. The symbols represent the calculated capacitance, and the solid lines are the corresponding linear fit, which show perfect linear fit with R2 = 1

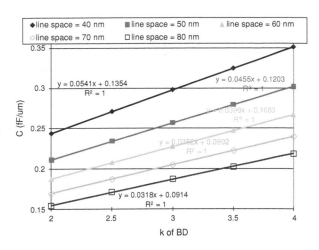

Fig. 6.14 The gradient dC/dk is plotted in terms of finger space and finger width

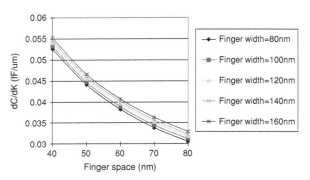

SOLID122 is a 3D 20-Node Electrostatic Solid and SOLID123 is a 3D 10-Node Tetrahedral Electrostatic Solid.

In a dual-damascene structure, misalignment sometimes is observed because the via and trench patterns are manufactured separately. In addition, surrounding corner structure is a common situation in a circuit layout where a metal line terminal is surrounded by another metal line as illustrated in Fig. 6.16 [11]. Using finite element analysis, it is found that significant electric field enhancement can be obtained for the two situations depicted in Fig. 6.16. At the surrounded metal-lead corners, a 2X increase in the nominal electric field was observed. For the misaligned via case, a 2.5X increase was determined. This suggests that for a 3.3 V application with 0.18 μm space, the local electric field can reach ∼0.5 MV/cm, and this is a significantly high electric field when one considers the breakdown strength of some of the low-k dielectrics can be relatively low (<2 MV/cm). The widely used 'E-model' for dielectric time-dependent-dielectric breakdown (TDDB) depicts that the time-to-failure (*TTF*) degrades exponentially on the electric field as follows [12]:

$$TTF = A * \exp(-\gamma E) * \exp(E_a/K_b T) \qquad (6.10)$$

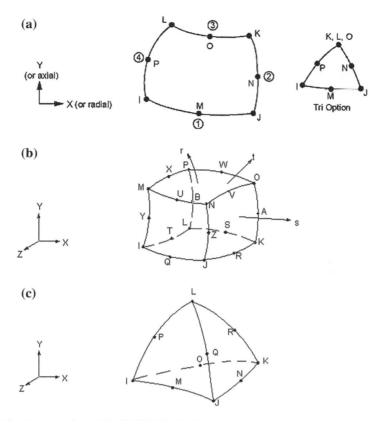

Fig. 6.15 Geometry for **a** 2-D PLANE121, **b** 3-D SOLID122 and **c** 3-D SOLID 123 [10]

Fig. 6.16 Electric field
enhancement by structural
effects [11], Copyright
© 2000, IEEE

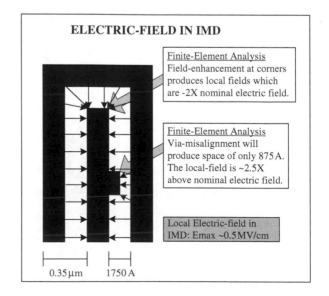

where γ is the field acceleration factor, which is typical ~ 4.1 cm/MV for low-k dielectrics [13], and E_a is the activation energy, T is the temperature and K_b is the Boltzmann constant. Therefore, the *TTF* under the 2X electric field can be estimated as follows:

$$\frac{TTF_{\text{high-field}}}{TTF_{\text{nominal}}} = \exp\left(-\gamma(E_{\text{high-field}} - E_{\text{nominal}})\right) \tag{6.11}$$

where the nominal electric field E_{nominal} is assumed to be 0.5 MV/cm, the $TTF_{\text{high-field}}$ will be only 0.12 of TTF_{nominal}. In other words, 2X electric field will lead to $\sim 88\%$ time-to-failure reduction.

Another excellent example of the FEM application to dielectric reliability is given by Chen et al. [14]. They performed the FEM simulation to examine the electric field distribution inside a multi-level VNCAP (vertical natural capacitors) as shown in Fig. 6.17a. VNCAP is a stacked metal/via structure which is characterized as low-cost, high density, and highly symmetric configurations. The electric field distribution for such multilevel structure should be much more complex than that shown in Fig. 6.16, and it cannot be explicitly obtained from any analytical analysis. The electric field distribution at each level depends not

Fig. 6.17 a 3D schematic of VNCAP structure containing four thin wire levels (M1–M4) which are interconnected by vias and one fat wire level (M5) without comb-via connection. **b** Cross section of VNCAP showing a typical dielectric breakdown event which preferentially took place at the M4 level [14], Copyright © 2006, IEEE

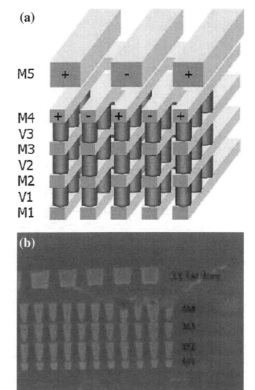

Fig. 6.18 FEM simulation of the electric field for a VNCAP with staggered via configuration. The field strengths for metal levels M3 and M4 were calculated along the dotted lines at the copper/cap and the low-*k*/cap interface, respectively [14], Copyright © 2006, IEEE

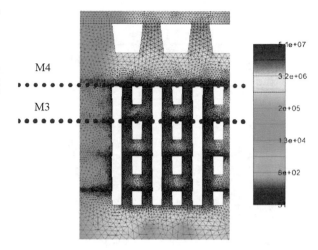

only on the surrounding via/metal environment, but also on the actual cap layer, thickness and materials. FEM was employed to investigate the root cause of the dominant dielectric breakdown at the M4 level as observed in Fig. 6.17b after TDDB testing. It is generally believed that there is some correlation between a shorter TDDB lifetime and a higher local electric field which results in faster Cu ion migration.

The finite element mesh of the structure in Fig 6.17 is shown in Fig. 6.18. Finer mesh was used around features with small curvatures because electric field concentration is expected there, as is commonly done. A 3.6 V bias was applied for the electric field analysis. Figure 6.19 shows that for both M3 and M4 layers, there are electric field spikes around the metal corners caused by corner effect as expected. However, the magnitude of the electric field spikes for M4 is about twice of those for M3, which should be attributed to the structural configuration difference between them. It is noted that M3 is of a more symmetrical characteristic in vertical direction since M3 is connected to both the upper layer (M4) and the lower layer (M2) through vias. However, top of M4 is covered by dielectric and no vias are connected. Therefore, field enhancement and thus a higher electric field are expected for M4.

In comparison to M3 Chen et al. [14] also found in Fig. 6.19 that the local electric field in M4 is slightly reduced at the low-*k*/cap interface but enhanced by about 7% at the copper/cap interface inside the copper line. They attributed this observation to the thicker cap layer on top of M4. As illustrated in Fig. 6.20, the local field enhancement inside the copper line is strongly influenced by the cap layer thickness, and it increases with increasing cap thickness. However, the field in the middle of the low-*k* (outside the copper line) is almost invariable, or even slightly decreases with increasing cap thickness.

Based on their experimental and simulation results, they hypothesized that the higher electric field at the Cu/cap interface inside the M4 Cu line could push more Cu ions towards the location where the spike electric field was, and they can be

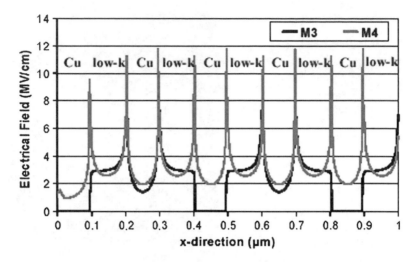

Fig. 6.19 FEM computed electric field distribution in M3 and M4 of the VNCAP structure along the dotted lines in Fig. 6.18 [14], Copyright © 2006, IEEE

Fig. 6.20 FEM simulations of the E-field for metal lines with three different cap layer thicknesses. The field strengths are calculated at two locations of the cap interface: (1) at the middle of the copper line; (2) at the middle of the low-*k* interface between the two metal lines [14], Copyright © 2006, IEEE

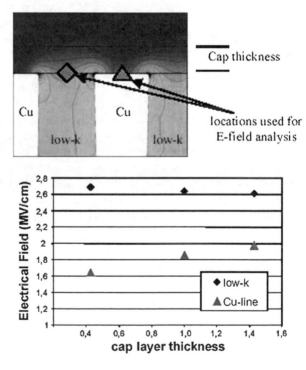

further injected into the dielectric/cap interface over the barrier. It is this electric field that controls the number of Cu ions injected along the interface. The higher electric field spike and the higher local E-field inside the copper line provide an explanation as to why M4 level was more vulnerable to breakdown than the other thin metal levels (M1, M2, and M3) with the same line width and space. Although the electric field at low-*k*/cap interface (which is expected to play a major role during copper ion diffusion process) is lower for M4, its impact to the TDDB lifetime is less pronounced since the number of injected copper ions (available copper ions participating in diffusion) at M4 is higher than other thin wire levels (M1, M2, and M3).

In summary, FEM provides insightful analysis to the dielectric reliability, especially for prevailing complex interconnect structures. FEM not only assists the electric field analysis, but also provides possible solution path for the dielectric reliability issues.

References

1. Semiconductor Industry Association (2005) International Technology Roadmap for Semiconductors. Semiconductor Industry Association, San Jose
2. Swanson Analysis System (1999) ANSYS Theroy Reference, 5.6 edn. Swanson Analysis System, Inc. (now ANSYS Inc.), Canosburg, USA
3. Swanson Analysis system (1999) ANSYS Element Reference, 5.6 edn. Swanson Analysis system, Inc. (now ANSYS Inc.), Canosburg, USA
4. Chang WH (1976) Analytical IC metal-line capacitance formulas. IEEE Trans Microw Theory Tech MTT-25:608–611
5. Chikaki S, Shimoyama M, Yagi R, Yoshino T, Shishida Y, Ono T, Ishikawa A, Fujii N, Nakayama T, Kohmura K, Tanaka H, Kawahara J, Matsuo H, Takada S, Yamanishi T, Hishiya S, Hata N, Kinoshita K, Kikkawa T (2005) Extraction of process-induced damage in low-k/Cu damascene structure. In: IEEE International Symposium on Semiconductor Manufacturing (ISSM), pp 422–425
6. Maex K, Baklanov MR, Shamiryan D, Iacopi F, Brongersma SH, Yanovitskaya ZS (2003) Low dielectric constant materials for microelectronics. J Appl Phys 93:8793–8841
7. http://www.electroline.com.au/elc/feature_article/item_042003a.asp. Accessed 21 May 2006
8. Kim CU. Available: http://www.sematech.org/meetings/past/20031027/TRC%202003_03_Kim.pdf. Accessed May 2006
9. Choudhury U, Sangiovanni-Vincentelli A (1995) Automatic generation of analytical models for interconnect capacitances. IEEE Trans Computer-Aided Des Integr Circuits Syst 14:470–480
10. http://www.kxcad.net/ansys/ANSYS/ansyshelp/. Accessed 22 Jan 2009
11. Tsu R, McPherson JW, McKee WR (2000) Leakage and breakdown reliability issues associated with low-k dielectrics in a dual-damascene Cu process. In: IEEE 38th Annual International Reliability Physics Symposium, San Jose, California, pp 348–353
12. Noguchi J (2005) Dominant factors in TDDB degradation of Cu interconnects. IEEE Trans Electron Devices 52:1743–1750
13. Haase GS, Ogawa ET, McPherson JW (2005) Breakdown characteristics of interconnect dielectrics. In: IEEE 43 Annual International Reliability Physics Symposium, pp 466–473
14. Chen F, Ungar F, Fischer AH, Gill J, Chinthakindi A, Goebel T, Shinosky M, Coolbaugh D, Ramachandran V, Siew YK, Kaltalioglu E, Kim SO, Park K (2006) Reliability characterization of BEOL vertical natural capacitor using copper and low-k SiCOH dielectric for 65 nm RF and mixed-signal applications. In: IEEE 44th Annual International Reliability Physics Symposium, San Jose, pp 490–495

Index

Printed by Publishers' Graphics LLC USA
MO20120404-035
2012